Numerical Solution of Partial Differential Equations

This is the second edition of a highly successful and well respected textbook on the numerical techniques used to solve partial differential equations arising from mathematical models in science, engineering and other fields.

Based upon the extensive experience of both authors, the book concentrates on the thorough treatment of standard numerical techniques which are chosen on the basis of their general utility for practical problems. As in the first edition, the emphasis is on finite difference methods for simple examples of parabolic, hyperbolic and elliptic equations, but there are brief descriptions of the finite element method and a new section on finite volume methods. Stability is treated clearly and rigorously by means of maximum principles, a completely revised section on energy methods, and discrete Fourier analysis. There are also new sections on modified equation analysis, symplectic integration schemes, convection–diffusion problems, multigrid methods and conjugate gradient methods.

Already an excellent choice for students and teachers in mathematics, engineering and computer science departments, the revised text brings the reader up to date with the latest developments in the subject.

Numerical Solution of Partial Differential Equations

An Introduction

K. W. Morton
University of Bath, UK

and

D. F. Mayers
University of Oxford, UK

Second Edition

CAMBRIDGE UNIVERSITY PRESS
Cambridge, New York, Melbourne, Madrid, Cape Town,
Singapore, São Paulo, Delhi, Tokyo, Mexico City

Cambridge University Press
The Edinburgh Building, Cambridge CB2 8RU, UK

Published in the United States of America by
Cambridge University Press, New York

www.cambridge.org
Information on this title: www.cambridge.org/9780521607933

First published 2005
7th printing 2011

A catalogue record for this publication is available from the British Library

ISBN 978-0-521-60793-3 Paperback

Contents

Preface to the first edition

The origin of this book was a sixteen-lecture course that each of us has given over the last several years to final-year Oxford undergraduate mathematicians; and its development owes much to the suggestions of some of our colleagues that the subject matter could readily be taught somewhat earlier as a companion course to one introducing the theory of partial differential equations. On the other hand, we have used much of the same material in teaching a one-year Master's course on mathematical modelling and numerical analysis. These two influences have guided our choice of topics and the level and manner of presentation.

Thus we concentrate on finite difference methods and their application to standard model problems. This allows the methods to be couched in simple terms while at the same time treating such concepts as stability and convergence with a reasonable degree of mathematical rigour. In a more advanced text, or one with greater emphasis on the finite element method, it would have been natural and convenient to use standard Sobolev space norms. We have avoided this temptation and used only discrete norms, specifically the maximum and the l_2 norms. There are several reasons for this decision. Firstly, of course, it is consistent with an aim of demanding the minimum in prerequisites – of analysis, of PDE theory, or of computing – so allowing the book to be used as a text in an early undergraduate course and for teaching scientists and engineers as well as mathematicians.

Equally importantly though, the decision fits in with our widespread use of discrete maximum principles in analysing methods for elliptic and parabolic problems, our treatment of discrete energy methods and conservation principles, and the study of discrete Fourier modes on finite domains. We believe that treating all these ideas at a purely discrete level helps to strengthen the student's understanding of these important mathematical tools. At the same time this is a very practical approach, and it encourages the interpretation of difference schemes as direct models of physical principles and phenomena: all calculations are, after all, carried out on a finite grid, and practical computations are checked for stability, etc. at the discrete level. Moreover, interpreting a difference scheme's effect on the Fourier modes that can be represented on the mesh, in terms of the damping and dispersion in one time step is often of greater value than considering the truncation error, which exemplifies the second justification of our approach.

However, the limiting process as a typical mesh parameter h tends to zero is vital to a proper understanding of numerical methods for partial differential equations. For example, if U^n is a discrete approximation at time level n and evolution through a time step Δt is represented as $U^{n+1} = C_h U^n$, many students find great difficulty in distinguishing the limiting process when $n \to \infty$ on a fixed mesh and with fixed Δt, from that in which $n \to \infty$ with $n\Delta t$ fixed and $h, \Delta t \to 0$. Both processes are of great practical importance: the former is related to the many iterative procedures that have been developed for solving the discrete equations approximating steady state problems by using the analogy of time stepping the unsteady problem; and understanding the latter is crucial to avoiding instability when choosing methods for approximating the unsteady problems themselves. The notions of uniform bounds and uniform convergence lie at the heart of the matter; and, of course, it is easier to deal with these by using norms which do not themselves depend on h. However, as shown for example by Palencia and Sanz–Serna,[1] a rigorous theory can be based on the use of discrete norms and this lies behind the approach we have adopted. It means that concepts such as well-posedness have to be rather carefully defined; but we believe the slight awkwardness entailed here is more than compensated for by the practical and pedagogical advantages pointed out above.

The ordering of the topics is deliberate and reflects the above concerns. We start with parabolic problems, which are both the simplest to approximate and analyse and also of widest utility. Through the addition of convection to the diffusion operator, this leads naturally to the study of hyperbolic problems. It is only after both these cases have been explored in some detail that, in Chapter 5, we present a careful treatment of the concepts of consistency, convergence and stability for evolutionary problems. The final two chapters are devoted respectively

[1] Palencia, C. and Sanz–Serna, J. M. (1984), An extension of the Lax–Richtmyer theory, *Numer. Math.* **44** (2), 279–283.

to the discretisation of elliptic problems, with a brief introduction to finite element methods, and to the iterative solution of the resulting algebraic equations; with the strong relationship between the latter and the solution of parabolic problems, the loop of linked topics is complete. In all cases, we present only a small number of methods, each of which is thoroughly analysed and whose practical utility we can attest to. Indeed, we have taken as a theme for the book that all the model problems and the methods used to approximate them are simple but generic.

Exercises of varying difficulty are given at the end of each chapter; they complete, extend or merely illustrate the text. They are all analytical in character, so the whole book could be used for a course which is purely theoretical. However, numerical analysis has very practical objectives, so there are many numerical illustrations of the methods given in the text; and further numerical experiments can readily be generated for students by following these examples. Computing facilities and practices develop so rapidly that we believe this open-ended approach is preferable to giving explicit practical exercises.

We have referred to the relevant literature in two ways. Where key ideas are introduced in the text and they are associated with specific original publications, full references are given in footnotes – as earlier in this Preface. In addition, at the end of each chapter we have included a brief section entitled 'Bibliographic notes and recommended reading' and the accumulated references are given at the end of the book. Neither of these sets of references is intended to be comprehensive, but they should enable interested students to pursue further their studies of the subject. We have, of course, been greatly guided and influenced by the treatment of evolutionary problems in Richtmyer and Morton (1967); in a sense the present book can be regarded as both introductory to and complementary to that text.

We are grateful to several of our colleagues for reading and commenting on early versions of the book (with Endre Süli's remarks being particularly helpful) and to many of our students for checking the exercises. The care and patience of our secretaries Jane Ellory and Joan Himpson over the long period of the book's development have above all made its completion possible.

Preface to the second edition

In the ten years since the first edition of this book was published, the numerical solution of PDEs has moved forward in many ways. But when we sought views on the main changes that should be made for this second edition, the general response was that we should not change the main thrust of the book or make very substantial changes. We therefore aimed to limit ourselves to adding no more than 10%–20% of new material and removing rather little of the original text: in the event, the book has increased by some 23%.

Finite difference methods remain the starting point for introducing most people to the solution of PDEs, both theoretically and as a tool for solving practical problems. So they still form the core of the book. But of course finite element methods dominate the elliptic equation scene, and finite volume methods are the preferred approach to the approximation of many hyperbolic problems. Moreover, the latter formulation also forms a valuable bridge between the two main methodologies. Thus we have introduced a new section on this topic in Chapter 4; and this has also enabled us to reinterpret standard difference schemes such as the Lax–Wendroff method and the box scheme in this way, and hence for example show how they are simply extended to nonuniform meshes. In addition, the finite element section in Chapter 6 has been followed by a new section on convection–diffusion problems: this covers both finite difference and finite element schemes and leads to the introduction of Petrov–Galerkin methods.

The theoretical framework for finite difference methods has been well established now for some time and has needed little revision. However, over the last few years there has been greater interaction between methods to approximate ODEs and those for PDEs, and we have responded to this stimulus in several ways. Firstly, the growing interest in applying symplectic methods to Hamiltonian ODE systems, and extending the approach to PDEs, has led to our including a section on this topic in Chapter 4 and applying the ideas to the analysis of the staggered leap–frog scheme used to approximate the system wave equation. More generally, the revived interest in the method of lines approach has prompted a complete redraft of the section on the energy method of stability analysis in Chapter 5, with important improvements in overall coherence as well as in the analysis of particular cases. In that chapter, too, is a new section on modified equation analysis: this technique was introduced thirty years ago, but improved interpretations of the approach for such as the box scheme have encouraged a reassessment of its position; moreover, it is again the case that its use for ODE approximations has both led to a strengthening of its analysis and a wider appreciation of its importance.

Much greater changes to our field have occurred in the practical application of the methods we have described. And, as we continue to have as our aim that the methods presented should properly represent and introduce what is used in practice, we have tried to reflect these changes in this new edition. In particular, there has been a huge improvement in methods for the iterative solution of large systems of algebraic equations. This has led to a much greater use of implicit methods for time-dependent problems, the widespread replacement of direct methods by iterative methods in finite element modelling of elliptic problems, and a closer interaction between the methods used for the two problem types. The emphasis of Chapter 7 has therefore been changed and two major sections added. These introduce the key topics of multigrid methods and conjugate gradient methods, which have together been largely responsible for these changes in practical computations.

We gave serious consideration to the possibility of including a number of MATLAB programs implementing and illustrating some of the key methods. However, when we considered how very much more powerful both personal computers and their software have become over the last ten years, we realised that such material would soon be considered outmoded and have therefore left this aspect of the book unchanged. We have also dealt with references to the literature and bibliographic notes in the same way as in the earlier edition: however, we have collected both into the reference list at the end of the book.

Solutions to the exercises at the end of each chapter are available in the form of LaTeX files. Those involved in teaching courses in this area can obtain copies, by email only, by applying to solutions@cambridge.org.

We are grateful to all those readers who have informed us of errors in the first edition. We hope we have corrected all of these and not introduced too many new ones. Once again we are grateful to our colleagues for reading and commenting on the new material.

1
Introduction

Partial differential equations (PDEs) form the basis of very many mathematical models of physical, chemical and biological phenomena, and more recently their use has spread into economics, financial forecasting, image processing and other fields. To investigate the predictions of PDE models of such phenomena it is often necessary to approximate their solution numerically, commonly in combination with the analysis of simple special cases; while in some of the recent instances the numerical models play an almost independent role.

Let us consider the design of an aircraft wing as shown in Fig. 1.1, though several other examples would have served our purpose equally well – such as the prediction of the weather, the effectiveness of pollutant dispersal, the design of a jet engine or an internal combustion engine,

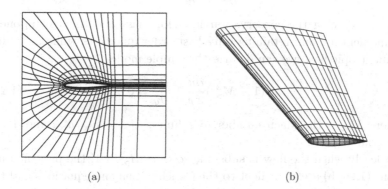

(a)   (b)

Fig. 1.1. (a) A typical (inviscid) computational mesh around an aerofoil cross-section; (b) a corresponding mesh on a wing surface.

the safety of a nuclear reactor, the exploration for and exploitation of oil, and so on.

In steady flight, two important design factors for a wing are the lift generated and the drag that is felt as a result of the flow of air past the wing. In calculating these quantities for a proposed design we know from boundary layer theory that, to a good approximation, there is a thin boundary layer near the wing surface where viscous forces are important and that outside this an inviscid flow can be assumed. Thus near the wing, which we will assume is locally flat, we can model the flow by

$$u\frac{\partial u}{\partial x} - \nu\frac{\partial^2 u}{\partial y^2} = (1/\rho)\frac{\partial p}{\partial x}, \qquad (1.1)$$

where u is the flow velocity in the direction of the tangential co-ordinate x, y is the normal co-ordinate, ν is the viscosity, ρ is the density and p the pressure; we have here neglected the normal velocity. This is a typical *parabolic* equation for u with $(1/\rho)\partial p/\partial x$ treated as a forcing term.

Away from the wing, considered just as a two-dimensional cross-section, we can suppose the flow velocity to be inviscid and of the form $(u_\infty + u, v)$ where u and v are small compared with the flow speed at infinity, u_∞ in the x-direction. One can often assume that the flow is irrotational so that we have

$$\frac{\partial v}{\partial x} - \frac{\partial u}{\partial y} = 0; \qquad (1.2a)$$

then combining the conservation laws for mass and the x-component of momentum, and retaining only first order quantities while assuming homentropic flow, we can deduce the simple model

$$(1 - M_\infty^2)\frac{\partial u}{\partial x} + \frac{\partial v}{\partial y} = 0 \qquad (1.2b)$$

where M_∞ is the Mach number at infinity, $M_\infty = u_\infty/a_\infty$, and a_∞ is the sound speed.

Clearly when the flow is subsonic so that $M_\infty < 1$, the pair of equations (1.2a, b) are equivalent to the Cauchy–Riemann equations and the system is *elliptic*. On the other hand for supersonic flow where $M_\infty > 1$, the system is equivalent to the one-dimensional wave equation and the system is *hyperbolic*. Alternatively, if we operate on (1.2b) by $\partial/\partial x$ and eliminate v by operating on (1.2a) by $\partial/\partial y$, we either obtain an equivalent to Laplace's equation or the second order wave equation.

Thus from this one situation we have extracted the three basic types of partial differential equation: we could equally well have done so from the other problem examples mentioned at the beginning. We know from PDE theory that the analysis of these three types, what constitutes a well-posed problem, what boundary conditions should be imposed and the nature of possible solutions, all differ very markedly. This is also true of their numerical solution and analysis.

In this book we shall concentrate on model problems of these three types because their understanding is fundamental to that of many more complicated systems. We shall consider methods, mainly finite difference methods and closely related finite volume methods, which can be used for more practical, complicated problems, but can only be analysed as thoroughly as is necessary in simpler situations. In this way we will be able to develop a rigorous analytical theory of such phenomena as stability and convergence when finite difference meshes are refined. Similarly, we can study in detail the speed of convergence of iterative methods for solving the systems of algebraic equations generated by difference methods. And the results will be broadly applicable to practical situations where precise analysis is not possible.

Although our emphasis will be on these separate equation types, we must emphasise that in many practical situations they occur together, in a system of equations. An example, which arises in very many applications, is the Euler–Poisson system: in two space dimensions and time t, they involve the two components of velocity and the pressure already introduced; then, using the more compact notation ∂_t for $\partial/\partial t$ etc., they take the form

$$\partial_t u + u\partial_x u + v\partial_y u + \partial_x p = 0$$
$$\partial_t v + u\partial_x v + v\partial_y v + \partial_y p = 0$$
$$\partial_x^2 p + \partial_y^2 p = 0. \tag{1.3}$$

Solving this system requires the combination of two very different techniques: for the final elliptic equation for p one needs to use the techniques described in Chapters 6 and 7 to solve a large system of simultaneous algebraic equations; then its solution provides the driving force for the first two hyperbolic equations, which can generally be solved by marching forward in time using techniques described in Chapters 2 to 5. Such a model typically arises when flow speeds are much lower than in aerodynamics, such as flow in a porous medium, like groundwater flow. The two procedures need to be closely integrated to be effective and efficient.

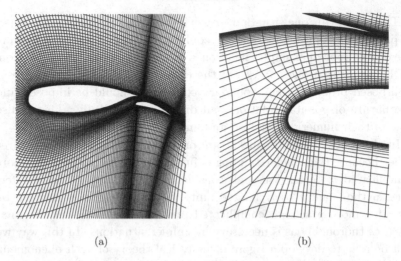

(a)   (b)

Fig. 1.2. A typical multi-aerofoil: (a) a general view; (b) a
detail of the mesh that might be needed for a Navier–Stokes
calculation. (Courtesy of DRA, Farnborough.)

Returning to our wing design example, however, it will be as well to
mention some of the practical complications that may arise. For a civil
aircraft most consideration can be given to its behaviour in steady flight
at its design speed; but, especially for a military aircraft, manoeuvrability
is important, which means that the flow will be unsteady and the equations time-dependent. Then, even for subsonic flow, the equations corresponding to (1.2a, b) will be hyperbolic (in one time and two space variables), similar to but more complicated than the Euler–Poisson system
(1.3). Greater geometric complexity must also be taken into account:
the three-dimensional form of the wing must be taken into consideration
particularly for the flow near the tip and the junction with the aircraft
body; and at landing and take-off, the flaps are extended to give greater
lift at the slower speeds, so in cross-section it may appear as in Fig. 1.2.

In addition, rather than the smooth flow regimes which we have so
far implicitly assumed, one needs in practice to study such phenomena as shocks, vortex sheets, turbulent wakes and their interactions.
Developments of the methods we shall study are used to model all
these situations but such topics are well beyond the scope of this book.
Present capabilities within the industry include the solution of approximations to the Reynolds-averaged Navier–Stokes equations for unsteady
viscous flow around a complete aircraft, such as that shown in Fig. 1.3.

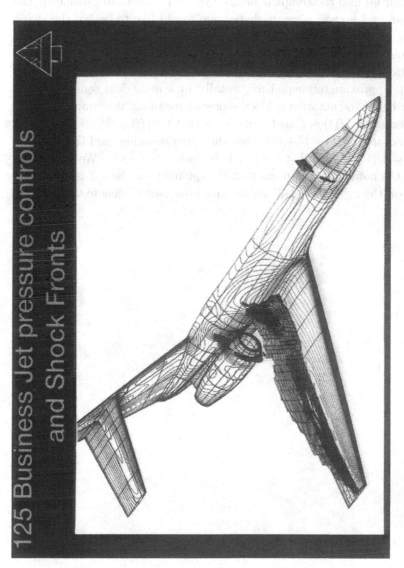

Fig. 1.3. Pressure contours and parts of the mesh for a complete aircraft at cruise conditions. (Courtesy of British Aerospace.)

Moreover, the ultimate objective is to integrate these flow prediction capabilities into the complete design cycle – rather than calculating the flow around a given aircraft shape, one would like to design the shape to obtain a given flow.

Finally, to end this introductory chapter there are a few points of notation to draw to the reader's attention. We use the notation $\approx$ to mean 'approximately equal to', usually in a numerical sense. On the other hand, the notation $\sim$ has the precise meaning 'is asymptotic to' in the sense that $f(t) \sim t^2$ as $t \to 0$ means that $t^{-2}[f(t) - t^2] \to 0$ as $t \to 0$. The notation $f(t) = t^2 + o(t^2)$ has the same meaning; and the notation $f(t) = O(t^2)$ means that $t^{-2}f(t)$ is bounded as $t \to 0$. We have often used the notation $:=$ to mean that the quantity on the left is *defined* by that on the right. We shall usually use bold face to denote vectors.

2
Parabolic equations in one space variable

2.1 Introduction

In this chapter we shall be concerned with the numerical solution of parabolic equations in one space variable and the time variable t. We begin with the simplest model problem, for heat conduction in a uniform medium. For this model problem an explicit difference method is very straightforward in use, and the analysis of its error is easily accomplished by the use of a maximum principle, or by Fourier analysis. As we shall show, however, the numerical solution becomes unstable unless the time step is severely restricted, so we shall go on to consider other, more elaborate, numerical methods which can avoid such a restriction. The additional complication in the numerical calculation is more than offset by the smaller number of time steps needed. We then extend the methods to problems with more general boundary conditions, then to more general linear parabolic equations. Finally we shall discuss the more difficult problem of the solution of nonlinear equations.

2.2 A model problem

Many problems in science and engineering are modelled by special cases of the linear parabolic equation for the unknown $u(x,t)$

$$\frac{\partial u}{\partial t} = \frac{\partial}{\partial x}\left(b(x,t)\frac{\partial u}{\partial x} \right) + c(x,t)u + d(x,t) \tag{2.1}$$

where b is strictly positive. An *initial condition* will be needed; if this is given at $t = 0$ it will take the form

$$u(x,0) = u^0(x) \tag{2.2}$$

7

where $u^0(x)$ is a given function. The solution of the problem will be required to satisfy (2.1) for $t > 0$ and x in an open region R which will be typically either the whole real line, the half-line $x > 0$, or an interval such as $(0, 1)$. In the two latter cases we require the solution to be defined on the closure of R and to satisfy certain *boundary conditions*; we shall assume that these also are linear, and may involve u or its first space derivative $\partial u/\partial x$, or both. If $x = 0$ is a left-hand boundary, the boundary condition will be of the form

$$\alpha_0(t)u + \alpha_1(t)\frac{\partial u}{\partial x} = \alpha_2(t) \tag{2.3}$$

where

$$\alpha_0 \geq 0, \ \alpha_1 \leq 0 \quad \text{and} \quad \alpha_0 - \alpha_1 > 0. \tag{2.4}$$

If $x = 1$ is a right-hand boundary we shall need a condition of the form

$$\beta_0(t)u + \beta_1(t)\frac{\partial u}{\partial x} = \beta_2(t) \tag{2.5}$$

where

$$\beta_0 \geq 0, \ \beta_1 \geq 0 \quad \text{and} \quad \beta_0 + \beta_1 > 0. \tag{2.6}$$

The reason for the conditions on the coefficients α and β will become apparent later. Note the change of sign between α_1 and β_1, reflecting the fact that at the right-hand boundary $\partial/\partial x$ is an outward normal derivative, while in (2.3) it was an inward derivative.

We shall begin by considering a simple model problem, the equation for which models the flow of heat in a homogeneous unchanging medium, of finite extent, with no heat source. We suppose that we are given homogeneous *Dirichlet boundary conditions*, i.e., the solution is given to be zero at each end of the range, for all values of t. After changing to dimensionless variables this problem becomes: find $u(x, t)$ defined for $x \in [0, 1]$ and $t \geq 0$ such that

$$u_t = u_{xx} \ \text{for} \ t > 0, \ 0 < x < 1, \tag{2.7}$$

$$u(0, t) = u(1, t) = 0 \ \text{for} \ t > 0, \tag{2.8}$$

$$u(x, 0) = u^0(x), \ \text{for} \ 0 \leq x \leq 1. \tag{2.9}$$

Here we have introduced the common subscript notation to denote partial derivatives.

2.3 Series approximation

This differential equation has special solutions which can be found by the method of *separation of variables*. The method is rather restricted in its application, unlike the finite difference methods which will be our main concern. However, it gives useful solutions for comparison purposes, and leads to a natural analysis of the stability of finite difference methods by the use of Fourier analysis.

We look for a solution of the special form $u(x,t) = f(x)g(t)$; substituting into the differential equation we obtain

$$fg' = f''g,$$

i.e.,
$$g'/g = f''/f. \tag{2.10}$$

In this last equation the left-hand side is independent of x, and the right-hand side is independent of t, so that both sides must be constant. Writing this constant as $-k^2$, we immediately solve two simple equations for the functions f and g, leading to the solution

$$u(x,t) = e^{-k^2 t} \sin kx.$$

This shows the reason for the choice of $-k^2$ for the constant; if we had chosen a positive value here, the solution would have involved an exponentially increasing function of t, whereas the solution of our model problem is known to be bounded for all positive values of t. For all values of the number k this is a solution of the differential equation; if we now restrict k to take the values $k = m\pi$, where m is a positive integer, the solution vanishes at $x = 1$ as well as at $x = 0$. Hence any linear combination of such solutions will satisfy the differential equation and the two boundary conditions. This linear combination can be written

$$u(x,t) = \sum_{m=1}^{\infty} a_m e^{-(m\pi)^2 t} \sin m\pi x. \tag{2.11}$$

We must now choose the coefficients a_m in this linear combination in order to satisfy the given initial condition. Writing $t = 0$ we obtain

$$\sum_{m=1}^{\infty} a_m \sin m\pi x = u^0(x). \tag{2.12}$$

This shows at once that the a_m are just the coefficients in the Fourier sine series expansion of the given function $u^0(x)$, and are therefore given by

$$a_m = 2 \int_0^1 u^0(x) \sin m\pi x \, dx. \tag{2.13}$$

This final result may be regarded as an exact analytic solution of the problem, but it is much more like a numerical approximation, for two reasons. If we require the value of $u(x,t)$ for specific values of x and t, we must first determine the Fourier coefficients a_m; these can be found exactly only for specially simple functions $u^0(x)$, and more generally would require some form of numerical integration. And secondly we can only sum a finite number of terms of the infinite series. For the model problem, however, it is a very efficient method; for even quite small values of t a few terms of the series will be quite sufficient, as the series converges extremely rapidly. The real limitation of the method in this form is that it does not easily generalise to even slightly more complicated differential equations.

2.4 An explicit scheme for the model problem

To approximate the model equation (2.7) by finite differences we divide the closed domain $\bar{R} \times [0, t_F]$ by a set of lines parallel to the x- and t-axes to form a grid or mesh. We shall assume, for simplicity only, that the sets of lines are equally spaced, and from now on we shall assume that $\bar{R}$ is the interval $[0, 1]$. Note that in practice we have to work in a finite time interval $[0, t_F]$, but t_F can be as large as we like.

We shall write Δx and Δt for the line spacings. The crossing points

$$(x_j = j\Delta x, \ t_n = n\Delta t), \ j = 0, 1, \ldots, J, \ n = 0, 1, \ldots, \tag{2.14}$$

where

$$\Delta x = 1/J, \tag{2.15}$$

are called the *grid points* or *mesh points*. We seek approximations of the solution at these mesh points; these approximate values will be denoted by

$$U_j^n \approx u(x_j, t_n). \tag{2.16}$$

We shall approximate the derivatives in (2.7) by finite differences and then solve the resulting difference equations in an evolutionary manner starting from $n = 0$.

We shall often use notation like U_j^n; there should be no confusion with other expressions which may look similar, such as λ^n which, of course, denotes the nth power of λ. If there is likely to be any ambiguity we shall sometimes write such a power in the form $(\lambda_j)^n$.

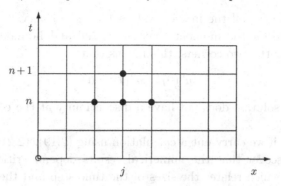

Fig. 2.1. An explicit scheme.

For the model problem the simplest difference scheme based at the mesh point (x_j, t_n) uses a forward difference for the time derivative; this gives

$$\frac{v(x_j, t_{n+1}) - v(x_j, t_n)}{\Delta t} \approx \frac{\partial v}{\partial t}(x_j, t_n) \qquad (2.17)$$

for any function v with a continuous t-derivative. The scheme uses a centred second difference for the second order space derivative:

$$\frac{v(x_{j+1}, t_n) - 2v(x_j, t_n) + v(x_{j-1}, t_n)}{(\Delta x)^2} \approx \frac{\partial^2 v}{\partial x^2}(x_j, t_n). \qquad (2.18)$$

The approximation generated by equating the left-hand sides of (2.17) and (2.18) thus satisfies

$$U_j^{n+1} = U_j^n + \mu(U_{j+1}^n - 2U_j^n + U_{j-1}^n) \qquad (2.19)$$

where

$$\mu := \frac{\Delta t}{(\Delta x)^2}. \qquad (2.20)$$

The pattern of grid points involved in (2.19) is shown in Fig. 2.1; clearly each value at time level t_{n+1} can be independently calculated from values at time level t_n; for this reason this is called an *explicit difference scheme*. From the initial and boundary values

$$U_j^0 = u^0(x_j), \; j = 1, 2, \ldots, J - 1, \qquad (2.21)$$

$$U_0^n = U_J^n = 0, \; n = 0, 1, 2, \ldots, \qquad (2.22)$$

we can calculate all the interior values for successive values of n. We shall assume for the moment that the initial and boundary data are consistent at the two corners; this means that

$$u^0(0) = u^0(1) = 0 \qquad (2.23)$$

so that the solution does not have a discontinuity at the corners of the domain.

However, if we carry out a calculation using (2.19), (2.21) and (2.22) we soon discover that the numerical results depend critically on the value of μ, which relates the sizes of the time step and the space step. In Fig. 2.2 we show results corresponding to initial data in the form of a 'hat function',

$$u^0(x) = \begin{cases} 2x & \text{if } 0 \le x \le \frac{1}{2}, \\ 2 - 2x & \text{if } \frac{1}{2} \le x \le 1. \end{cases} \qquad (2.24)$$

Two sets of results are displayed; both use $J = 20$, $\Delta x = 0.05$. The first set uses $\Delta t = 0.0012$, and the second uses $\Delta t = 0.0013$. The former clearly gives quite an accurate result, while the latter exhibits oscillations which grow rapidly with increasing values of t. This is a typical example of *stability* or *instability* depending on the value of the mesh ratio μ. The difference between the behaviour of the two numerical solutions is quite striking; these solutions use time steps which are very nearly equal, but different enough to give quite different forms of numerical solution.

We shall now analyse this behaviour, and obtain bounds on the error, in a more formal way. First we introduce some notation and definitions.

2.5 Difference notation and truncation error

We define finite differences in the same way in the two variables t and x; there are three kinds of finite differences:

forward differences

$$\Delta_{+t} v(x, t) := v(x, t + \Delta t) - v(x, t), \qquad (2.25a)$$
$$\Delta_{+x} v(x, t) := v(x + \Delta x, t) - v(x, t); \qquad (2.25b)$$

backward differences

$$\Delta_{-t} v(x, t) := v(x, t) - v(x, t - \Delta t), \qquad (2.26a)$$
$$\Delta_{-x} v(x, t) := v(x, t) - v(x - \Delta x, t); \qquad (2.26b)$$

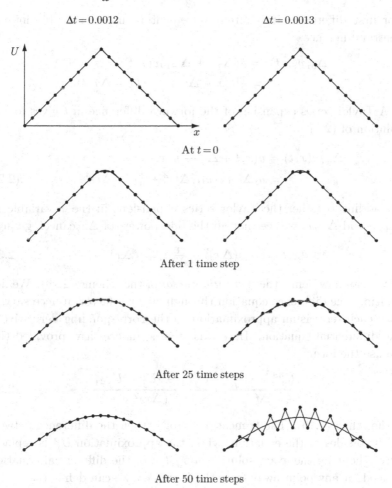

Fig. 2.2. Results obtained for the data of (2.24) with the explicit method; $J = 20$, $\Delta x = 0.05$. The exact solution is shown by the full curved line.

central differences

$$\delta_t v(x, t) := v(x, t + \tfrac{1}{2}\Delta t) - v(x, t - \tfrac{1}{2}\Delta t), \qquad (2.27\text{a})$$
$$\delta_x v(x, t) := v(x + \tfrac{1}{2}\Delta x, t) - v(x - \tfrac{1}{2}\Delta x, t). \qquad (2.27\text{b})$$

When the central difference operator is applied twice we obtain the very useful second order central difference

$$\delta_x^2 v(x, t) := v(x + \Delta x, t) - 2v(x, t) + v(x - \Delta x, t). \qquad (2.28)$$

For first differences it is often convenient to use the double interval central difference

$$\Delta_{0x}v(x,t) := \tfrac{1}{2}(\Delta_{+x} + \Delta_{-x})v(x,t)$$
$$= \tfrac{1}{2}[v(x + \Delta x, t) - v(x - \Delta x, t)].$$

A Taylor series expansion of the forward difference in t gives for the solution of (2.7)

$$\Delta_{+t}u(x,t) = u(x, t + \Delta t) - u(x,t)$$
$$= u_t\Delta t + \tfrac{1}{2}u_{tt}(\Delta t)^2 + \tfrac{1}{6}u_{ttt}(\Delta t)^3 + \cdots . \qquad (2.29)$$

By adding together the Taylor series expansions in the x variable for $\Delta_{+x}u$ and $\Delta_{-x}u$, we see that all the odd powers of Δx cancel, giving

$$\delta_x^2\, u(x,t) = u_{xx}(\Delta x)^2 + \tfrac{1}{12}u_{xxxx}(\Delta x)^4 + \cdots . \qquad (2.30)$$

We can now define the *truncation error* of the scheme (2.19). We first multiply the difference equation throughout by a factor, if necessary, so that each term is an approximation to the corresponding derivative in the differential equation. Here this step is unnecessary, provided that we use the form

$$\frac{U_j^{n+1} - U_j^n}{\Delta t} = \frac{U_{j+1}^n - 2U_j^n + U_{j-1}^n}{(\Delta x)^2} \qquad (2.31)$$

rather than (2.19). The truncation error is then the difference between the two sides of the equation, when the approximation U_j^n is replaced throughout by the exact solution $u(x_j, t_n)$ of the differential equation. Indeed, at any point away from the boundary we can define the

truncation error $T(x,t)$

$$T(x,t) := \frac{\Delta_{+t}u(x,t)}{\Delta t} - \frac{\delta_x^2 u(x,t)}{(\Delta x)^2} \qquad (2.32)$$

so that

$$T(x,t) = (u_t - u_{xx}) + \left(\tfrac{1}{2}u_{tt}\Delta t - \tfrac{1}{12}u_{xxxx}(\Delta x)^2\right) + \cdots$$
$$= \tfrac{1}{2}u_{tt}\Delta t - \tfrac{1}{12}u_{xxxx}(\Delta x)^2 + \cdots \qquad (2.33)$$

where these leading terms are called the *principal part* of the truncation error, and we have used the fact that u satisfies the differential equation.

We have used Taylor series expansions to express the truncation error as an infinite series. It is often convenient to truncate the infinite Taylor series, introducing a remainder term; for example

$$u(x, t + \Delta t) = u(x,t) + u_t \Delta t + \tfrac{1}{2} u_{tt} (\Delta t)^2 + \tfrac{1}{6} u_{ttt} (\Delta t)^3 + \cdots$$
$$= u(x,t) + u_t \Delta t + \tfrac{1}{2} u_{tt}(x, \eta)(\Delta t)^2, \tag{2.34}$$

where η lies somewhere between t and $t + \Delta t$. If we do the same thing for the x expansion the truncation error becomes

$$T(x,t) = \tfrac{1}{2} u_{tt}(x, \eta) \Delta t - \tfrac{1}{12} u_{xxxx}(\xi, t)(\Delta x)^2 \tag{2.35}$$

where $\xi \in (x - \Delta x, x + \Delta x)$, from which it follows that

$$|T(x,t)| \le \tfrac{1}{2} M_{tt} \Delta t + \tfrac{1}{12} M_{xxxx}(\Delta x)^2 \tag{2.36}$$
$$= \tfrac{1}{2} \Delta t \left[M_{tt} + \tfrac{1}{6\mu} M_{xxxx} \right], \tag{2.37}$$

where M_{tt} is a bound for $|u_{tt}|$ and M_{xxxx} is a bound for $|u_{xxxx}|$. It is now clear why we assumed that the initial and boundary data for u were consistent, and why it is helpful if we can also assume that the initial data are sufficiently smooth. For then we can assume that the bounds M_{tt} and M_{xxxx} hold uniformly over the closed domain $[0, 1] \times [0, t_F]$. Otherwise we must rely on the smoothing effect of the diffusion operator to ensure that for any $\tau > 0$ we can find bounds of this form which hold for the domain $[0, 1] \times [\tau, t_F]$. This sort of difficulty can easily arise in problems which look quite straightforward. For example, suppose the boundary conditions specify that u must vanish on the boundaries $x = 0$ and $x = 1$, and that u must take the value 1 on the initial line, where $t = 0$. Then the solution $u(x,t)$ is obviously discontinuous at the corners, and in the full domain defined by $0 < x < 1$, $t > 0$ all its derivatives are unbounded, so our bound for the truncation error is useless over the full domain. We shall see later how this problem can be treated by Fourier analysis.

For the problem of Fig. 2.2 we see that

$$T(x,t) \to 0 \quad \text{as} \quad \Delta x, \ \Delta t \ \to \ 0 \ \ \forall (x,t) \in (0, 1) \times [\tau, t_F),$$

independently of any relation between the two mesh sizes. We say that the scheme is *unconditionally consistent* with the differential equation. For a fixed ratio μ we also see from (2.37) that $|T|$ will behave asymptotically like $O(\Delta t)$ as $\Delta t \to 0$: except for special values of μ this will be the highest power of Δt for which such a statement could be made, so that the scheme is said to have *first order accuracy*.

However, it is worth noting here that, since u satisfies $u_t = u_{xx}$ everywhere, we also have $u_{tt} = u_{xxxx}$ and hence

$$T(x,t) = \tfrac{1}{2}\left(1 - \tfrac{1}{6\mu}\right)u_{xxxx}\Delta t + O\big((\Delta t)^2\big).$$

Thus for $\mu = \tfrac{1}{6}$ the scheme is *second order accurate*. This however is a rather special case. Not only does it apply just for this particular choice of μ, but also for more general equations with variable coefficients it cannot hold. For example, in the solution of the equation $u_t = b(x,t)u_{xx}$ it would require choosing a different time step Δt at each point.

2.6 Convergence of the explicit scheme

Now suppose that we carry out a sequence of calculations using the same initial data, and the same value of $\mu = \Delta t/(\Delta x)^2$, but with successive refinement of the two meshes, so that $\Delta t \to 0$ and $\Delta x \to 0$. Then we say that the scheme is *convergent* if, for any fixed point (x^*, t^*) in a given domain $(0,1) \times (\tau, t_F)$,

$$x_j \to x^*, \quad t_n \to t^* \quad \text{implies} \quad U_j^n \to u(x^*, t^*). \tag{2.38}$$

We shall prove that the explicit scheme for our problem is convergent if $\mu \le \tfrac{1}{2}$.

We need consider only points (x^*, t^*) which coincide with mesh points for sufficiently refined meshes; for convergence at all other points will follow from the continuity of $u(x,t)$. We also suppose that we can introduce an upper bound $\bar{T} = \bar{T}(\Delta x, \Delta t)$ for the truncation error, which holds for all mesh points on a given mesh, and use the notation T_j^n for $T(x_j, t_n)$:

$$|T_j^n| \le \bar{T}. \tag{2.39}$$

We denote by e the error $U - u$ in the approximation; more precisely

$$e_j^n := U_j^n - u(x_j, t_n). \tag{2.40}$$

Now U_j^n satisfies the equation (2.19) exactly, while $u(x_j, t_n)$ leaves the remainder $T_j^n \Delta t$; this follows immediately from the definition of T_j^n. Hence by subtraction we obtain

$$e_j^{n+1} = e_j^n + \mu\delta_x^2 e_j^n - T_j^n\Delta t \tag{2.41}$$

which is in detail

$$e_j^{n+1} = (1 - 2\mu)e_j^n + \mu e_{j+1}^n + \mu e_{j-1}^n - T_j^n\Delta t. \tag{2.42}$$

The important point for the proof is that if $\mu \leq \frac{1}{2}$ the coefficients of the three terms e^n on the right of this equation are all positive, and add up to unity. If we introduce the maximum error at a time step by writing

$$E^n := \max\{|e_j^n|, \quad j = 0, 1, \ldots, J\}, \tag{2.43}$$

the fact that the coefficients are positive means that we can omit the modulus signs in the triangle inequality to give

$$|e_j^{n+1}| \leq (1 - 2\mu)E^n + \mu E^n + \mu E^n + |T_j^n|\Delta t$$
$$\leq E^n + \bar{T}\Delta t. \tag{2.44}$$

Since this inequality holds for all values of j from 1 to $J - 1$, we have

$$E^{n+1} \leq E^n + \bar{T}\Delta t. \tag{2.45}$$

Suppose for the moment that the bound (2.39) holds on the finite interval $[0, t_F]$; and since we are using the given initial values for U_j^n we know that $E^0 = 0$. A very simple induction argument then shows that $E^n \leq n\bar{T}\Delta t$. Hence we obtain from (2.37)

$$E^n \leq \frac{1}{2}\Delta t \left[M_{tt} + \frac{1}{6\mu}M_{xxxx} \right] t_F$$
$$\to 0 \quad \text{as} \quad \Delta t \to 0. \tag{2.46}$$

In our model problem, if it is useful we can write $M_{tt} = M_{xxxx}$.

We can now state this convergence property in slightly more general terms. In order to define convergence of a difference scheme which involves two mesh sizes Δt and Δx we need to be clear about what relationship we assume between them as they both tend to zero. We therefore introduce the concept of a refinement path. A *refinement path* is a sequence of pairs of mesh sizes, Δx and Δt, each of which tends to zero:

$$refinement\ path := \{((\Delta x)_i, (\Delta t)_i), i = 0, 1, 2, \ldots; (\Delta x)_i, (\Delta t)_i \to 0\}. \tag{2.47}$$

We can then specify particular refinement paths by requiring, for example, that $(\Delta t)_i$ is proportional to $(\Delta x)_i$, or to $(\Delta x)_i^2$. Here we just define

$$\mu_i = \frac{(\Delta t)_i}{(\Delta x)_i^2} \tag{2.48}$$

and merely require that $\mu_i \leq \frac{1}{2}$. Some examples are shown in Fig. 2.3.

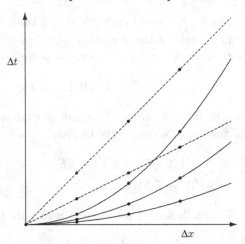

Fig. 2.3. Refinement paths; shown as full lines for constant $\Delta t/(\Delta x)^2$ and as dashed lines for constant $\Delta t/\Delta x$.

Theorem 2.1 *If a refinement path satisfies $\mu_i \leq \frac{1}{2}$ for all sufficiently large values of i, and the positive numbers n_i, j_i are such that*

$$n_i(\Delta t)_i \rightarrow t > 0, \quad j_i(\Delta x)_i \rightarrow x \in [0,1],$$

and if $|u_{xxxx}| \leq M_{xxxx}$ uniformly in $[0,1] \times [0, t_F]$, then the approximations $U_{j_i}^{n_i}$ generated by the explicit difference scheme (2.19) for $i = 0, 1, 2 \ldots$ converge to the solution $u(x,t)$ of the differential equation, uniformly in the region.

Such a convergence theorem is the least that one can expect of a numerical scheme; it shows that arbitrarily high accuracy can be attained by use of a sufficiently fine mesh. Of course, it is also somewhat impractical. As the mesh becomes finer, more and more steps of calculation are required, and the effect of rounding errors in the calculation would become significant and would eventually completely swamp the truncation error.

As an example with smoother properties than is given by the data of (2.24), consider the solution of the heat equation with

$$u(x,0) = x(1-x), \tag{2.49a}$$

$$u(0,t) = u(1,t) = 0, \tag{2.49b}$$

on the region $[0,1] \times [0,1]$. Errors obtained with the explicit method are shown in Fig. 2.4. This shows a graph of $\log_{10} E^n$ against t_n, where

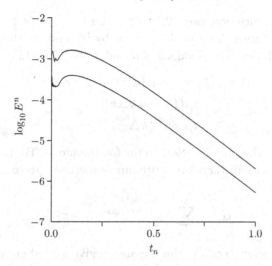

Fig. 2.4. Error decay for the explicit method applied to the heat equation with initial condition $u(x, 0) = x(1 - x)$. The top curve is for $\Delta x = 0.1$, $\mu = 0.5$, and the bottom curve is for $\Delta x = 0.05$, $\mu = 0.5$.

E^n is given by (2.43). Two curves are shown; one uses $J = 10$, $\Delta x = 0.1$, and the other uses $J = 20$, $\Delta x = 0.05$. Both have $\mu = \frac{1}{2}$, which is the largest value consistent with stability. The two curves show clearly how the error behaves as the grid size is reduced: they are very similar in shape, and for each value of t_n the ratio of the two values of E^n is close to 4, the ratio of the values of $\Delta t = \frac{1}{2}(\Delta x)^2$. Notice also that after some early variation the error tends to zero as t increases; our error bound in (2.45) is pessimistic, as it continues to increase with t. The early variation in the error results from the lack of smoothness in the corners of the domain already referred to. We will discuss this in more detail in the next section and in Section 2.10.

2.7 Fourier analysis of the error

We have already expressed the exact solution of the differential equation as a Fourier series; this expression is based on the observation that a particular set of Fourier modes are exact solutions. We can now easily show that a similar Fourier mode is an exact solution of the difference equations. Suppose we substitute

$$U_j^n = (\lambda)^n e^{ik(j\Delta x)} \tag{2.50}$$

into the difference equation (2.19), putting $U_j^{n+1} = \lambda U_j^n$ and similarly for the other terms. We can then divide by U_j^n and see that this Fourier mode is a solution for all values of n and j provided that

$$\lambda \equiv \lambda(k) = 1 + \mu \left(e^{ik\Delta x} - 2 + e^{-ik\Delta x} \right)$$
$$= 1 - 2\mu(1 - \cos k\Delta x)$$
$$= 1 - 4\mu \sin^2 \tfrac{1}{2} k\Delta x; \qquad (2.51)$$

$\lambda(k)$ is called the *amplification factor* for the mode. By taking $k = m\pi$ as in (2.11), we can therefore write our numerical approximation in the form

$$U_j^n = \sum_{-\infty}^{\infty} A_m e^{im\pi(j\Delta x)} \left[\lambda(m\pi) \right]^n . \qquad (2.52)$$

The low frequency terms in this expansion give a good approximation to the exact solution of the differential equation, given by (2.11), because the series expansions for $\lambda(k)$ and $\exp(-k^2 \Delta t)$ match reasonably well:

$$\exp(-k^2 \Delta t) = 1 - k^2 \Delta t + \tfrac{1}{2} k^4 (\Delta t)^2 - \cdots ,$$
$$\lambda(k) = 1 - 2\mu \left[\tfrac{1}{2} (k\Delta x)^2 - \tfrac{1}{24} (k\Delta x)^4 + \cdots \right]$$
$$= 1 - k^2 \Delta t + \tfrac{1}{12} k^4 \Delta t (\Delta x)^2 - \cdots . \qquad (2.53)$$

Indeed these expansions provide an alternative means of investigating the truncation error of our scheme. It is easy to see that we will have at least first order accuracy, but when $(\Delta x)^2 = 6\Delta t$ we shall have second order accuracy. In fact it is quite easy to show that there exists a constant $C(\mu)$ depending only on the value of μ such that

$$|\lambda(k) - e^{-k^2 \Delta t}| \le C(\mu) k^4 (\Delta t)^2 \qquad \forall k, \Delta t > 0. \qquad (2.54)$$

Theorem 2.1 establishes convergence and an error bound under the restriction $\mu \le \tfrac{1}{2}$, but it does not show what happens if this condition is not satisfied. Our analysis of the Fourier modes shows what happens to the high frequency components in this case. For large values of k the modes in the exact solution are rapidly damped by the exponential term $\exp(-k^2 t)$. But in the numerical solution the damping factor $|\lambda(k)|$ will become greater than unity for large values of k if $\mu > \tfrac{1}{2}$; in particular this will happen when $k\Delta x = \pi$, for then $\lambda(k) = 1 - 4\mu$. These Fourier modes will then grow unboundedly as n increases. It is possible in principle to choose the initial conditions so that these Fourier modes do not appear in the solution. But this would be a very special problem, and in practice the effect of rounding errors would be to introduce small components of all the modes, some of which would then grow without bound. For the present model problem we shall say that a method is *stable* if there exists a constant K, independent of k, such that

$$|[\lambda(k)]^n| \leq K, \quad \text{for} \quad n\Delta t \leq t_F, \forall k. \tag{2.55}$$

Essentially, stability has to do with the bounded growth in a finite time of the difference between two solutions of the difference equations, uniformly in the mesh size; we shall formulate a general definition of stability in a later chapter. Evidently, for stability we require the condition, due to von Neumann,

$$|\lambda(k)| \leq 1 + K'\Delta t \tag{2.56}$$

to hold for all k. We shall find that such a stability condition is necessary and sufficient for the convergence of a consistent difference scheme approximating a single differential equation. Thus for the present model problem the method is unstable when $\mu > \frac{1}{2}$ and stable when $\mu \leq \frac{1}{2}$.

We have used a representation for U_j^n as the infinite Fourier series (2.52), since it is easily comparable with the exact solution. However on the discrete mesh there are only a finite number of distinct modes; modes with wave numbers k_1 and k_2 are indistinguishable if $(k_1 - k_2)\Delta x$ is a multiple of 2π. It may therefore be more convenient to expand U_j^n as a linear combination of the distinct modes corresponding to

$$k = m\pi, \quad m = -(J-1), -(J-2), \ldots, -1, 0, 1, \ldots, J. \tag{2.57}$$

The highest mode which can be carried by the mesh has $k = J\pi$, or $k\Delta x = \pi$; this mode has the values ± 1 at alternate points on the mesh. We see from (2.51) that it is also the most unstable mode for this difference scheme, as it often is for many difference schemes, and has the amplification factor $\lambda(J\pi) = 1 - 4\mu$. It is the fastest growing mode when $\mu > \frac{1}{2}$, which is why it eventually dominates the solutions shown in Fig. 2.2.

We can also use this Fourier analysis to extend the convergence theorem to the case where the initial data $u^0(x)$ are continuous on $[0,1]$, but may not be smooth, in particular at the corners. We no longer have to assume that the solution has sufficient bounded derivatives that u_{xxxx} and u_{tt} are uniformly bounded on the region considered. Instead we just assume that the Fourier series expansion of $u^0(x)$ is absolutely convergent. We suppose that μ is fixed, and that $\mu \leq \frac{1}{2}$. Consider the error, as before,

$$
\begin{aligned}
e_j^n &= U_j^n - u(x_j, t_n) \\
&= \sum_{-\infty}^{\infty} A_m e^{im\pi j \Delta x} \left\{ [\lambda(m\pi)]^n - e^{-m^2\pi^2 n\Delta t} \right\}, \quad (2.58)
\end{aligned}
$$

where we have also used the full Fourier series for $u(x, t)$ instead of the sine series as in the particular case of (2.11); this will allow treatment other than of the simple boundary conditions of (2.8). We now split this infinite sum into two parts. Given an arbitrary positive ϵ, we choose m_0 such that

$$
\sum_{|m|>m_0} |A_m| \leq \frac{1}{4}\epsilon. \quad (2.59)
$$

We know that this is possible, because of the absolute convergence of the series. If both $|\lambda_1| \leq 1$ and $|\lambda_2| \leq 1$, then

$$
|(\lambda_1)^n - (\lambda_2)^n| \leq n|\lambda_1 - \lambda_2|; \quad (2.60)
$$

so from (2.54) we have

$$
\begin{aligned}
|e_j^n| &\leq \frac{1}{2}\epsilon + \sum_{|m|\leq m_0} |A_m| \left| [\lambda(m\pi)]^n - e^{-m^2\pi^2 n\Delta t} \right| \\
&\leq \frac{1}{2}\epsilon + \sum_{|m|\leq m_0} |A_m| nC(\mu) \left(m^2\pi^2 \Delta t \right)^2. \quad (2.61)
\end{aligned}
$$

We can thus deduce that

$$
|e_j^n| \leq \frac{1}{2}\epsilon + t_F C(\mu)\pi^4 \left[\sum_{|m|\leq m_0} |A_m| m^4 \right] \Delta t \quad (2.62)
$$

and by taking Δt sufficiently small we can obtain $|e_j^n| \leq \epsilon$ for all (x_j, t_n) in $[0, 1] \times [0, t_F]$. Note how the sum involving $A_m m^4$ plays much the same role as the bound on u_{xxxx} in the earlier analysis, but by making more precise use of the stability properties of the scheme we do not require that this sum is convergent.

2.8 An implicit method

The stability limit $\Delta t \leq \frac{1}{2}(\Delta x)^2$ is a very severe restriction, and implies that very many time steps will be necessary to follow the solution over a reasonably large time interval. Moreover, if we need to reduce Δx

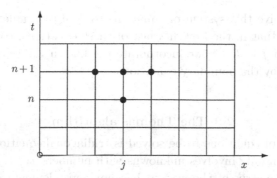

Fig. 2.5. The fully implicit scheme.

to improve the accuracy of the solution the amount of work involved increases very rapidly, since we shall also have to reduce Δt. We shall now show how the use of a backward time difference gives a difference scheme which avoids this restriction, but at the cost of a slightly more sophisticated calculation.

If we replace the forward time difference by the backward time difference, the space difference remaining the same, we obtain the scheme

$$\frac{U_j^{n+1} - U_j^n}{\Delta t} = \frac{U_{j+1}^{n+1} - 2U_j^{n+1} + U_{j-1}^{n+1}}{(\Delta x)^2} \qquad (2.63)$$

instead of (2.31). This may be written using the difference notation given in Section 2.5 as

$$\Delta_{-t} U_j^{n+1} = \mu \delta_x^2 U_j^{n+1},$$

where $\mu = \Delta t/(\Delta x)^2$, and has the stencil shown in Fig. 2.5.

This is an example of an *implicit* scheme, which is not so easy to use as the explicit scheme described earlier. The scheme (2.63) involves three unknown values of U on the new time level $n + 1$; we cannot immediately calculate the value of U_j^{n+1} since the equation involves the two neighbouring values U_{j+1}^{n+1} and U_{j-1}^{n+1}, which are also unknown. We must now write the equation in the form

$$-\mu U_{j-1}^{n+1} + (1 + 2\mu)U_j^{n+1} - \mu U_{j+1}^{n+1} = U_j^n. \qquad (2.64)$$

Giving j the values $1, 2, \ldots, (J - 1)$ we thus obtain a system of $J - 1$ linear equations in the $J - 1$ unknowns $U_j^{n+1}, j = 1, 2, \ldots, J - 1$. Instead of calculating each of these unknowns by a separate trivial formula, we must now solve this system of equations to give the values simultaneously. Note that in the first and last of these equations, corresponding to $j = 1$ and $j = J - 1$, we incorporate the known values of U_0^{n+1} and U_J^{n+1} given by the boundary conditions.

2.9 The Thomas algorithm

The system of equations to be solved is tridiagonal:equation number j in the system only involves unknowns with numbers $j - 1, j$ and $j + 1$, so that the matrix of the system has non-zero elements only on the diagonal and in the positions immediately to the left and to the right of the diagonal. We shall meet such systems again, and it is useful here to consider a more general system of the form

$$-a_j U_{j-1} + b_j U_j - c_j U_{j+1} = d_j, \quad j = 1, 2, \ldots, J - 1, \qquad (2.65)$$

with

$$U_0 = 0, \quad U_J = 0. \qquad (2.66)$$

Here we have written the unknowns U_j, omitting the superscript for the moment. The coefficients a_j, b_j and c_j, and the right-hand side d_j, are given, and we assume that they satisfy the conditions

$$a_j > 0, \quad b_j > 0, \quad c_j > 0, \qquad (2.67)$$

$$b_j > a_j + c_j. \qquad (2.68)$$

Though stronger than necessary, these conditions ensure that the matrix is *diagonally dominant*, with the diagonal element in each row being at least as large as the sum of the absolute values of the other elements. It is easy to see that these conditions are satisfied by our difference equation system.

The Thomas algorithm operates by reducing the system of equations to upper triangular form, by eliminating the term in U_{j-1} in each of the equations. This is done by treating each equation in turn. Suppose that the first k of equations (2.65) have been reduced to

$$U_j - e_j U_{j+1} = f_j, \quad j = 1, 2, \ldots, k. \qquad (2.69)$$

The last of these equations is therefore $U_k - e_k U_{k+1} = f_k$, and the next equation, which is still in its original form, is

$$-a_{k+1} U_k + b_{k+1} U_{k+1} - c_{k+1} U_{k+2} = d_{k+1}.$$

It is easy to eliminate U_k from these two equations, giving a new equation involving U_{k+1} and U_{k+2},

$$U_{k+1} - \frac{c_{k+1}}{b_{k+1} - a_{k+1}e_k}U_{k+2} = \frac{d_{k+1} + a_{k+1}f_k}{b_{k+1} - a_{k+1}e_k}.$$

Comparing this with (2.69) shows that the coefficients e_j and f_j can be obtained from the recurrence relations

$$e_j = \frac{c_j}{b_j - a_je_{j-1}}, \quad f_j = \frac{d_j + a_jf_{j-1}}{b_j - a_je_{j-1}}, \quad j = 1, 2, \ldots, J-1; \quad (2.70)$$

while identifying the boundary condition $U_0 = 0$ with (2.69) for $j = 0$ gives the initial values

$$e_0 = f_0 = 0. \tag{2.71}$$

Having used these recurrence relations to find the coefficients, the values of U_j are easily obtained from (2.69): beginning from the known value of U_J, this equation gives the values of $U_{J-1}, U_{J-2}, \ldots$, in order, finishing with U_1.

The use of a recurrence relation like (2.69) to calculate the values of U_j in succession may in general be numerically unstable, and lead to increasing errors. However, this will not happen if, for each j, $|e_j| < 1$ in (2.69), and we leave it as an exercise to show that the conditions (2.67) and (2.68) are sufficient to guarantee this (see Exercise 4).

The algorithm is very efficient (on a serial computer) so that (2.64) is solved with 3(add) + 3(multiply) + 2(divide) operations per mesh point, as compared with 2(add) + 2(multiply) operations per mesh point for the explicit algorithm (2.19). Thus it takes about twice as long for each time step. The importance of the implicit method is, of course, that the time steps can be much larger, for, as we shall see, there is no longer any stability restriction on Δt. We shall give a proof of the convergence of this implicit scheme in the next section, as a particular case of a more general method. First we can examine its stability by the Fourier method of Section 2.7.

We construct a solution of the difference equations for Fourier modes of the same form as before,

$$U_j^n = (\lambda)^n e^{ik(j\Delta x)}. \tag{2.72}$$

This will satisfy (2.64) provided that

$$\lambda - 1 = \mu\lambda(e^{ik\Delta x} - 2 + e^{-ik\Delta x})$$
$$= -4\mu\lambda \sin^2 \tfrac{1}{2}k\Delta x, \tag{2.73}$$

which shows that

$$\lambda = \frac{1}{1 + 4\mu \sin^2 \frac{1}{2}k\Delta x}. \tag{2.74}$$

Evidently we have $0 < \lambda < 1$ for any positive choice of μ, so that this implicit method is *unconditionally stable*. As we shall see in the next section, the truncation error is much the same size as that of the explicit scheme, but we no longer require any restriction on μ to ensure that no Fourier mode grows as n increases.

The time step is still limited by the requirement that the truncation error must stay small, but in practice it is found that in most problems the implicit method can use a much larger Δt than the explicit method; although each step takes about twice as much work, the overall amount of work required to reach the time t_F is much less.

2.10 The weighted average or θ-method

We have now considered two finite difference methods, which differ only in that one approximates the second space derivative by three points on the old time level, t_n, and the other uses the three points on the new time level, t_{n+1}. A natural generalisation is to an approximation which uses all six of these points. This can be regarded as taking a weighted average of the two formulae. Since the time difference on the left-hand sides is the same, we obtain the six-point scheme (see Fig. 2.6)

$$U_j^{n+1} - U_j^n = \mu \left[\theta \delta_x^2 U_j^{n+1} + (1 - \theta)\delta_x^2 U_j^n \right], \quad j = 1, 2, \ldots, J - 1. \tag{2.75}$$

We shall assume that we are using an average with nonnegative weights, so that $0 \le \theta \le 1$; $\theta = 0$ gives the explicit scheme, $\theta = 1$ the fully implicit scheme. For any $\theta \ne 0$, we have a tridiagonal system to solve for $\{U_j^{n+1}\}$, namely,

$$-\theta\mu U_{j-1}^{n+1} + (1 + 2\theta\mu)U_j^{n+1} - \theta\mu U_{j+1}^{n+1} = \left[1 + (1 - \theta)\mu\delta_x^2 \right] U_j^n. \tag{2.76}$$

Clearly the coefficients satisfy (2.67) and (2.68), so the system can be solved stably by the Thomas algorithm given above for the fully implicit scheme.

Let us consider the stability of this one-parameter family of schemes by using Fourier analysis as in Section 2.7 and above. Substituting the mode (2.72) into equation (2.75), we obtain

$$\lambda - 1 = \mu[\theta\lambda + (1 - \theta)] \left(e^{ik\Delta x} - 2 + e^{-ik\Delta x} \right)$$
$$= \mu[\theta\lambda + (1 - \theta)] \left(-4\sin^2 \frac{1}{2}k\Delta x \right),$$

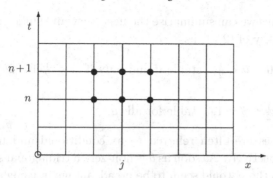

Fig. 2.6. The θ-method.

i.e.,

$$\lambda = \frac{1 - 4(1 - \theta)\mu \sin^2 \frac{1}{2}k\Delta x}{1 + 4\theta\mu \sin^2 \frac{1}{2}k\Delta x}. \tag{2.77}$$

Because $\mu > 0$, and we are assuming that $0 \leq \theta \leq 1$, it is clear that we can never have $\lambda > 1$: thus instability arises only through the possibility that $\lambda < -1$, that is that

$$1 - 4(1 - \theta)\mu \sin^2 \tfrac{1}{2}k\Delta x < - \left[1 + 4\theta\mu \sin^2 \tfrac{1}{2}k\Delta x\right],$$

i.e.,

$$4\mu(1 - 2\theta) \sin^2 \tfrac{1}{2}k\Delta x > 2.$$

The mode most liable to instability is the one for which the left side is largest: as before this is the most rapidly oscillatory mode, for which $k\Delta x = \pi$. This is an unstable mode if

$$\mu(1 - 2\theta) > \tfrac{1}{2}. \tag{2.78}$$

This includes the earlier explicit case, $\theta = 0$: and it also shows that the fully implicit scheme with $\theta = 1$ is not unstable for any value of μ. Indeed no scheme with $\theta \geq \frac{1}{2}$ is unstable for any μ. If condition (2.78) is satisfied there can be unbounded growth over a fixed time as $\Delta t \to 0$ and hence $n \to \infty$: on the other hand if (2.78) is not satisfied, we have $|\lambda(k)| \leq 1$ for every mode k, so that no mode grows at all and the scheme is stable. Thus we can summarise the necessary and sufficient conditions for the stability of (2.75) as

$$\left. \begin{array}{l} \text{when } 0 \le \theta < \tfrac{1}{2}, \quad \text{stable if and only if } \mu \le \tfrac{1}{2}(1 - 2\theta)^{-1} \\[1.5em] \text{when } \tfrac{1}{2} \le \theta \le 1, \quad \text{stable for all } \mu. \end{array} \right\} \quad (2.79)$$

The two cases are often referred to as conditional and unconditional stability respectively. As soon as θ is non-zero a tridiagonal system has to be solved, so there would seem to be no advantage in using schemes with $0 < \theta < \tfrac{1}{2}$ which are only conditionally stable – unless they were more accurate. Thus we should next look at the truncation error for (2.75).

To calculate the truncation error for such a six-point scheme it is important to make a careful choice of the point about which the Taylor series are to be expanded. It is clear that the leading terms in the truncation error will be the same for any choice of this expansion point: but the convenience and simplicity of the calculation can be very materially affected. Thus for the explicit scheme (2.31) the natural and convenient point was (x_j, t_n): by the same argument the natural point of expansion for the purely implicit scheme (2.63) would be (x_j, t_{n+1}). However, for any intermediate value of θ we shall use the centre of the six mesh points, namely $(x_j, t_{n+1/2})$, and often write the truncation error as $T_j^{n+1/2}$. It is also helpful to group the terms in the scheme in a symmetric manner so as to take maximum advantage of cancellations in the Taylor expansions. Working from (2.75) we therefore have, using the superscript/subscript notation for u as well as U,

$$u_j^{n+1} = \left[u + \tfrac{1}{2}\Delta t\, u_t + \tfrac{1}{2}\left(\tfrac{1}{2}\Delta t\right)^2 u_{tt} + \tfrac{1}{6}\left(\tfrac{1}{2}\Delta t\right)^3 u_{ttt} + \cdots \right]_j^{n+1/2},$$

$$u_j^n = \left[u - \tfrac{1}{2}\Delta t\, u_t + \tfrac{1}{2}\left(\tfrac{1}{2}\Delta t\right)^2 u_{tt} - \tfrac{1}{6}\left(\tfrac{1}{2}\Delta t\right)^3 u_{ttt} + \cdots \right]_j^{n+1/2}.$$

If we subtract these two series, all the even terms of the two Taylor series will cancel, and we obtain

$$\delta_t u_j^{n+1/2} = u_j^{n+1} - u_j^n = \left[\Delta t\, u_t + \tfrac{1}{24}(\Delta t)^3 u_{ttt} + \cdots \right]_j^{n+1/2}. \quad (2.80)$$

Also from (2.30) we have

$$\delta_x^2 u_j^{n+1} = \left[(\Delta x)^2 u_{xx} + \tfrac{1}{12}(\Delta x)^4 u_{xxxx} + \tfrac{2}{6!}(\Delta x)^6 u_{xxxxxx} + \cdots \right]_j^{n+1}. \quad (2.81)$$

We now expand each term in this series in powers of Δt, about the point $(x_j, t_{n+1/2})$. For simplicity in presenting these expansions, we omit the superscript and subscript, so it is understood that the resulting expressions are all to be evaluated at this point. This gives

$$\delta_x^2 u_j^{n+1} = \left[(\Delta x)^2 u_{xx} + \tfrac{1}{12}(\Delta x)^4 u_{xxxx} + \tfrac{2}{6!}(\Delta x)^6 u_{xxxxxx} + \cdots\right]$$
$$+ \tfrac{1}{2}\Delta t \left[(\Delta x)^2 u_{xxt} + \tfrac{1}{12}(\Delta x)^4 u_{xxxxt} + \cdots\right]$$
$$+ \tfrac{1}{2}\left(\tfrac{1}{2}\Delta t\right)^2 \left[(\Delta x)^2 u_{xxtt} + \cdots\right] + \cdots.$$

There is a similar expansion for $\delta_x^2 u_j^n$: combining these we obtain

$$\theta \delta_x^2 u_j^{n+1} + (1-\theta)\delta_x^2 u_j^n =$$
$$\left[(\Delta x)^2 u_{xx} + \tfrac{1}{12}(\Delta x)^4 u_{xxxx} + \tfrac{2}{6!}(\Delta x)^6 u_{xxxxxx} + \cdots\right]$$
$$+ (\theta - \tfrac{1}{2})\Delta t \left[(\Delta x)^2 u_{xxt} + \tfrac{1}{12}(\Delta x)^4 u_{xxxxt} + \cdots\right]$$
$$+ \tfrac{1}{8}(\Delta t)^2 (\Delta x)^2 \left[u_{xxtt}\right] + \cdots. \qquad (2.82)$$

Here we have retained more terms than we shall normally need to calculate the principal part of the truncation error, in order to show clearly the pattern for all the terms involved. In addition we have not exploited yet the fact that u is to satisfy the differential equation, so that (2.80) and (2.82) hold for any sufficiently smooth functions. If we now use these expansions to calculate the truncation error we obtain

$$T_j^{n+1/2} := \frac{\delta_t u_j^{n+1/2}}{\Delta t} - \frac{\theta \delta_x^2 u_j^{n+1} + (1-\theta)\delta_x^2 u_j^n}{(\Delta x)^2} \qquad (2.83)$$
$$= \left[u_t - u_{xx}\right] + \left[\left(\tfrac{1}{2} - \theta\right)\Delta t\, u_{xxt} - \tfrac{1}{12}(\Delta x)^2 u_{xxxx}\right]$$
$$+ \left[\tfrac{1}{24}(\Delta t)^2 u_{ttt} - \tfrac{1}{8}(\Delta t)^2 u_{xxtt}\right]$$
$$+ \left[\tfrac{1}{12}\left(\tfrac{1}{2} - \theta\right)\Delta t\,(\Delta x)^2 u_{xxxxt} - \tfrac{2}{6!}(\Delta x)^4 u_{xxxxxx}\right] \qquad (2.84)$$

where we have still not carried out any cancellations but have merely grouped terms which are ripe for cancellation.

The first term in (2.84) always cancels, so confirming consistency for all values of θ and μ . The second shows that we shall normally have first order accuracy (in Δt) but that the symmetric average $\theta = \tfrac{1}{2}$ is special: this value gives the well known and popular *Crank–Nicolson scheme*, named after those two authors who in a 1947 paper[1] applied the scheme very successfully to problems in the dyeing of textiles. Since the third

[1] Crank, J. and Nicolson, P. (1947) A practical method for numerical evaluation of solutions of partial differential equations of the heat-conduction type. *Proc. Camb. Philos. Soc.* **43**, 50–67.

term in (2.84) does not cancel even when we exploit the differential equation to obtain

$$T_j^{n+1/2} = -\tfrac{1}{12} \left[(\Delta x)^2 u_{xxxx} + (\Delta t)^2 u_{ttt} \right]_j^{n+1/2} + \cdots \quad (2.85)$$

(when $\theta = \tfrac{1}{2}$),

we see that the Crank–Nicolson scheme is always second order accurate in both Δt and Δx: this means that we can exploit the extra stability properties of the scheme to take larger time steps, with for example $\Delta x = O(\Delta t)$, and because then the truncation error is $O((\Delta t)^2)$ we can achieve good accuracy economically.

Another choice which is sometimes advocated is a generalisation of that discussed in Section 2.5. It involves eliminating the second term in (2.84) completely by relating the choice of θ to that of Δt and Δx so that

$$\theta = \tfrac{1}{2} - (\Delta x)^2 / 12 \Delta t, \quad (2.86)$$

i.e.,

$$\mu = \frac{1}{6(1 - 2\theta)}, \quad (2.87)$$

but note that this requires $(\Delta x)^2 \leq 6\Delta t$ to ensure $\theta \geq 0$. This gives a value of θ less than $\tfrac{1}{2}$ but it is easy to see that the condition (2.79) is satisfied, so that it is stable. It reduces to $\mu = \tfrac{1}{6}$ for the explicit case $\theta = 0$. The resulting truncation error is

$$T_j^{n+1/2} = -\tfrac{1}{12} \left[(\Delta t)^2 u_{ttt} + \tfrac{1}{20} (\Delta x)^4 u_{xxxxxx} \right]_j^{n+1/2} + \cdots \quad (2.88)$$

(when $\theta = \tfrac{1}{2} - \tfrac{1}{12\mu}$),

which is $O\left((\Delta t)^2 + (\Delta x)^4 \right)$. Thus again we can take large time steps while maintaining accuracy and stability: for example, with $\Delta t = \Delta x = 0.1$ we find we have $\theta = \tfrac{1}{2} - \tfrac{1}{120}$ so the scheme is quite close to the Crank–Nicolson scheme.

There are many other possible difference schemes that could be used for the heat flow equation and in Richtmyer and Morton (1967) (pp. 189–91), some fourteen schemes are tabulated. However the two-time-level, three-space-point schemes of (2.75) are by far the most widely used in practice, although the best choice of the parameter θ varies from problem to problem. Even for a given problem there may not be general agreement as to which scheme is the best. In the next section we consider the convergence analysis of these more general methods: but first we give

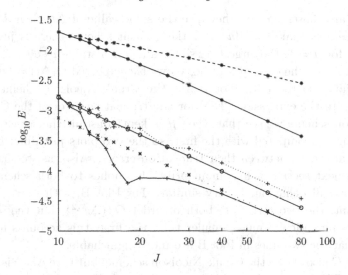

Fig. 2.7. Maximum error on $[0,1] \times [0.1,1]$ plotted against J, for various schemes.

$A:$ $\theta = 0,\ \mu = \frac{1}{2}$ —0—0—0—0

$B:$ $\theta = \frac{1}{2},\ \mu = \frac{1}{2},$ ×—×—×—×—

       $\theta = \frac{1}{2},\ \nu = \frac{1}{20}$ $\cdots$ × $\cdots$ × $\cdots$

$C:$ $\theta = \frac{1}{2},\ \mu = 5$ —+—+—+—

       $\theta = \frac{1}{2},\ \nu = \frac{1}{2}$ $\cdots$+$\cdots$+$\cdots$

$D:$ $\theta = 1,\ \mu = 5$ —*—*—*—

       $\theta = 1,\ \nu = \frac{1}{2}$ - - -*- - -*- - -*- - -

results for the problem of (2.49) obtained with implicit methods, which show similar behaviour to those of Fig. 2.4 obtained with the explicit method, with the Crank–Nicolson method being particularly accurate. In the set of graphs in Fig. 2.7 the maximum error E is plotted against the number of mesh points J for various schemes: to eliminate transient behaviour for small t we have used

$$E := \max\left\{\left|e_j^n\right|, (x_j, t_n) \in [0,1] \times [0.1,1]\right\}.$$

We start with $J = 10$; for each implicit scheme we show a graph with fixed $\mu = \Delta t/(\Delta x)^2$ as a solid line, and also a graph with fixed $\nu = \Delta t/\Delta x$ as a dotted line; note that in the latter case the number of time steps that are needed increases much more slowly. The values of μ

and ν are chosen so that they give the same value of Δt when $J = 10$; this requires that $\mu = 10\nu$. For the explicit scheme there is just one graph, for $\mu = \frac{1}{2}$, the largest possible value for a stable result.

Plot A, for the explicit scheme, shows the expected $O(\Delta t) = O(J^{-2})$ behaviour; so too does Plot B, for the Crank–Nicolson scheme with $\mu = \frac{1}{2}$. In the expression (2.85) for the truncation error of the Crank–Nicolson scheme we see that when μ is kept constant the second term is negligible compared with the first, so the two plots just differ by the fixed ratio of $\frac{1}{2}$ between their truncation errors. Also, as we shall see in the next section a maximum principle applies to both schemes so their overall behaviour is very similar. For Plot B, with $\nu = \frac{1}{20}$ kept constant, the two terms are both of order $O\left((\Delta x)^2\right)$, but the second term is numerically much smaller than the first; this accounts for the fact that the two lines in Plot B are indistinguishable.

Plot C also shows the Crank–Nicolson scheme, but here $\mu = 5$ is much larger. The numerical solution has oscillatory behaviour for small t, and the two graphs in Plot C are therefore much more erratic, not settling to their expected behaviour until J is about 40. For J larger than this the two graphs with $\mu = \frac{1}{2}$ and $\mu = 5$ are close together, illustrating the fact that the leading term in the truncation error in (2.85) is independent of μ. However, when $\nu = \frac{1}{2}$ is constant, the second term in (2.85) is a good deal larger than the first, and when $J > 40$ this graph lies well above the corresponding line in Plot B. Further analysis in Section 5.8 will help to explain this behaviour.

For the fully implicit method (plot D), where the maximum principle will apply again, the results are poor but as expected: with $\mu = 5$ we have $O(\Delta t) = O(J^{-2})$ behaviour; and with $\Delta t/\Delta x = \frac{1}{2}$ we get only $O(\Delta t) = O(J^{-1})$ error reduction.

These graphs do not give a true picture of the relative effectiveness of the various θ-methods because they do not take account of the work involved in each calculation. So in the graphs in Fig. 2.8 the same results are plotted against a measure of the computational effort involved in each calculation: for each method this should be roughly proportional to the total number of mesh points $(\Delta x \, \Delta t)^{-1}$, with the explicit method requiring approximately half the effort of the implicit methods. The two lines in Plot B are no longer the same: when J increases with fixed ν the time step Δt decreases more slowly than when μ is fixed, so less computational work is required. These graphs show that, for this problem, the Crank–Nicolson method with $\nu = \frac{1}{2}$ is the most efficient of those tested, provided that J is taken large enough to remove the initial

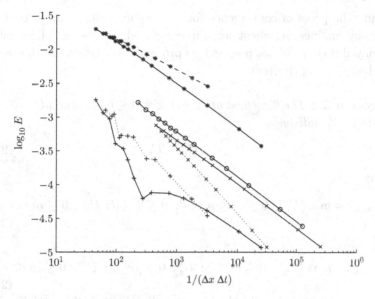

Fig. 2.8. Maximum error on $[0,1] \times [0.1,1]$ plotted against the total number of mesh points for various schemes.

$$
\begin{array}{lll}
A: & \theta = 0,\ \mu = \frac{1}{2} & \text{—0—0—0—0} \\
B: & \theta = \frac{1}{2},\ \mu = \frac{1}{2}, & \text{—x—x—x—} \\
 & \theta = \frac{1}{2},\ \nu = \frac{1}{20} & \text{$\cdots$x$\cdots$x$\cdots$} \\
C: & \theta = \frac{1}{2},\ \mu = 5 & \text{—+—+—+—} \\
 & \theta = \frac{1}{2},\ \nu = \frac{1}{2} & \text{$\cdots$+$\cdots$+$\cdots$} \\
D: & \theta = 1,\ \mu = 5 & \text{—*—*—*—} \\
 & \theta = 1,\ \nu = \frac{1}{2} & \text{- - -*- - -*- - -*- - -}
\end{array}
$$

oscillations; but a comparison with the $\nu = \frac{1}{20}$ plot suggests alternative choices of ν might be better.

2.11 A maximum principle and convergence for $\mu(1-\theta) \leq \frac{1}{2}$

If we consider what other properties a difference approximation to $u_t = u_{xx}$ should possess beyond convergence as $\Delta t, \Delta x \to 0$ (together with the necessary stability and a reasonable order of accuracy), a natural next requirement is a maximum principle. For we know mathematically (and by common experience if u represents, say, temperature) that $u(x,t)$ is bounded above and below by the extremes attained by the initial data and the values on the boundary up to time t. Such a principle also lay

behind the proof of convergence for the explicit scheme in Section 2.6:
and any engineering client for our computed results would be rather
dismayed if they did not possess this property. We generalise that result
by the following theorem.

Theorem 2.2 *The θ-method of (2.75) with $0 \le \theta \le 1$ and $\mu(1-\theta) \le \frac{1}{2}$
yields $\left\{U_j^n\right\}$ satisfying*

$$U_{min} \le U_j^n \le U_{max} \tag{2.89}$$

where

$$U_{min} := \min\left\{U_0^m,\ 0 \le m \le n;\ U_j^0,\ 0 \le j \le J;\ U_J^m,\ 0 \le m \le n\right\}, \tag{2.90}$$

and

$$U_{max} := \max\left\{U_0^m,\ 0 \le m \le n;\ U_j^0,\ 0 \le j \le J;\ U_J^m,\ 0 \le m \le n\right\}. \tag{2.91}$$

*For any refinement path which eventually satisfies this stability con-
dition, the approximations given by (2.75) with consistent initial and
Dirichlet boundary data converge uniformly on $[0,1] \times [0, t_F]$ if the initial data are smooth enough for the truncation error $T_j^{n+1/2}$ to tend to
zero along the refinement path uniformly in this domain.*

Proof We write (2.75) in the form

$$(1 + 2\theta\mu)U_j^{n+1} = \theta\mu\left(U_{j-1}^{n+1} + U_{j+1}^{n+1}\right) + (1-\theta)\mu\left(U_{j-1}^n + U_{j+1}^n\right)$$
$$+ \left[1 - 2(1-\theta)\mu\right]U_j^n. \tag{2.92}$$

Then under the hypotheses of the theorem all the coefficients on the right
are nonnegative and sum to $(1 + 2\theta\mu)$. Now suppose that U attains its
maximum at an internal point, and this maximum is U_j^{n+1}, and let U^*
be the greatest of the five values of U appearing on the right-hand side of
(2.92). Then since the coefficients are nonnegative $U_j^{n+1} \le U^*$; but since
this is assumed to be the maximum value, we also have $U_j^{n+1} \ge U^*$, so
$U_j^{n+1} = U^*$. Indeed, the maximum value must also be attained at each
neighbouring point which has a non-zero coefficient in (2.92). The same
argument can then be applied at each of these points, showing that the
maximum is attained at a sequence of points, until a boundary point
is reached. The maximum is therefore attained at a boundary point.
An identical argument shows that the minimum is also attained at a
boundary point, and the first part of the proof is complete.

By the definition of truncation error (see (2.84)), the solution of the differential equation satisfies the same relation as (2.92) except for an additional term $\Delta t\, T_j^{n+1/2}$ on the right-hand side. Thus the error $e_j^n = U_j^n - u_j^n$ is determined from the relations

$$(1 + 2\theta\mu)e_j^{n+1} = \theta\mu\left(e_{j-1}^{n+1} + e_{j+1}^{n+1}\right) + (1-\theta)\mu\left(e_{j-1}^n + e_{j+1}^n\right)$$
$$+\left[1 - 2(1-\theta)\mu\right]e_j^n - \Delta t T_j^{n+1/2} \qquad (2.93)$$

for $j = 1, 2, \ldots, J-1$ and $n = 0, 1, \ldots$ together with initial and boundary conditions. Suppose first of all that these latter are zero because $U_j^0 = u_j^0$, $U_0^m = u_0^m$ and $U_J^m = u_J^m$. Then we define, as in Section 2.6,

$$E^n := \max_{0 \le j \le J} \left|e_j^n\right|, \quad T^{n+1/2} := \max_{1 \le j \le J-1} \left|T_j^{n+1/2}\right|. \qquad (2.94)$$

Because of the nonnegative coefficients, it follows that

$$(1 + 2\theta\mu)E^{n+1} \le 2\theta\mu E^{n+1} + E^n + \Delta t\, T^{n+1/2}$$

and hence that

$$E^{n+1} \le E^n + \Delta t\, T^{n+1/2} \qquad (2.95)$$

so that, since $E^0 = 0$,

$$E^n \le \Delta t \sum_0^{n-1} T^{m+1/2},$$
$$\le n\Delta t \max_m T^{m+1/2} \qquad (2.96)$$

and this tends to zero along the refinement path under the assumed hypotheses.

So far we have assumed that numerical errors arise from the truncation errors of the finite difference approximations, but that the boundary values are used exactly. Suppose now that there are errors in the initial and boundary values of U_j^n and let us denote them by ϵ_j^0, ϵ_0^m and ϵ_J^m with $0 \le j \le J$ and $0 \le m \le N$, say. Then the errors e_j^n satisfy the recurrence relation (2.93) with initial and boundary values

$$e_j^0 = \epsilon_j^0, \quad j = 0, 1, \ldots, n,$$
$$e_0^m = \epsilon_0^m, \ e_J^m = \epsilon_J^m, \quad 0 \le m \le N.$$

Then (by Duhamel's principle) e_j^N can be written as the sum of two terms. The first term satisfies (2.93) with zero initial and boundary values; this term is bounded by (2.96). The second term satisfies the homogeneous form of (2.93), with the term in T omitted, and with the given non-zero initial and boundary values. By the maximum principle this term must lie between the maximum and minimum values of these initial and boundary values. Thus the error of the numerical solution will tend to zero along the refinement path, as required, provided that the initial and boundary values are *consistent*; that is, the errors in the initial and boundary values also tend to zero along the refinement path.

$\square$

The condition for this theorem, $\mu(1 - \theta) \leq \frac{1}{2}$, is very much more restrictive than that needed in the Fourier analysis of stability, $\mu(1 - 2\theta) \leq \frac{1}{2}$; for example, the Crank–Nicolson scheme always satisfies the stability condition, but only if $\mu \leq 1$ does it satisfy the condition given for a maximum principle, which in the theorem is then used to deduce stability and convergence. In view of this large gap the reader may wonder about the sharpness of this theorem. In fact, the maximum principle condition is sharp, but a little severe: for with $J = 2$ and $U_0^0 = U_2^0 = 0$, $U_1^0 = 1$, one obtains $U_1^1 = 1 - 2(1 - \theta)\mu$ which is nonnegative only if the given condition is satisfied; but, of course, one would use larger values of J in practice and this would relax the condition a little (see Exercise 11). Moreover, if with $U_0^n = U_J^n = 0$ one wants to have

$$|U_j^n| \leq K \max_{0 \leq i \leq J} |U_i^0| \quad \forall j, n \tag{2.97}$$

satisfied with $K = 1$, which is the property needed to deduce the error bound (2.96), it has recently been shown[1] that it is necessary and sufficient that $\mu(1 - \theta) \leq \frac{1}{4}(2 - \theta)/(1 - \theta)$, giving $\mu \leq \frac{3}{2}$ for the Crank–Nicolson scheme. It is only when any value of K is accepted in this growth bound, which is all that is required in the stability definition of (2.55), that the weaker condition $\mu(1 - 2\theta) \leq \frac{1}{2}$ is adequate. Then for Crank–Nicolson one can actually show that $K \leq 23$ holds!

Thus the maximum principle analysis can be viewed as an alternative means of obtaining stability conditions. It has the advantage over Fourier analysis that it is easily extended to apply to problems with variable coefficients (see below in Section 2.15); but, as we see above, it is easy to derive only sufficient stability conditions.

These points are illustrated in Fig. 2.9. Here the model problem is solved by the Crank–Nicolson scheme. The boundary conditions specify that the solution is zero at each end of the range, and the initial condition

[1] See Kraaijevanger, J.F.B.M. (1992) Maximum norm contractivity of discretization schemes for the heat equation. *Appl. Numer. Math.* **99**, 475–92.

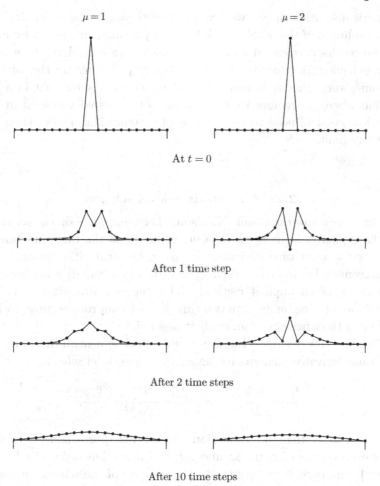

$\mu = 1$ $\mu = 2$

At $t = 0$

After 1 time step

After 2 time steps

After 10 time steps

Fig. 2.9. The Crank–Nicolson method applied to the heat equation where the initial distribution has a sharp spike at the mid-point; $J = 20$, $\Delta x = 0.05$.

gives the values of U_j^0 to be zero except at the mid-point; the value at the mid-point is unity. This corresponds to a function with a sharp spike at $x = \frac{1}{2}$.

In the case $\mu = 2$ the maximum principle does not hold, and we see that at the first time level the numerical solution becomes negative at the mid-point. This would normally be regarded as unacceptable. When $\mu = 1$ the maximum principle holds, and the numerical values all lie between 0 and 1, as required. However, at the first time level the numerical solution shows two peaks, one each side of the mid-point; the exact solution of the problem will have only a single maximum for all t. These results correspond to a rather extreme case, and the unacceptable behaviour only persists for a few time steps; thereafter the solution becomes very smooth in each case. However, they show that in a situation where we require to model some sort of rapid variation in the solution we shall need to use a value of μ somewhat smaller than the stability limit.

2.12 A three-time-level scheme

We have seen how the Crank–Nicolson scheme improves on the accuracy of the explicit scheme by the use of symmetry in the time direction to remove the even time derivative terms in the truncation error. This improvement has to be balanced against the extra complication involved in the use of an implicit method. This suggests investigation of the possibility of using more than two time levels to improve accuracy, while retaining the efficiency of an explicit method.

Consider, for example, the use of a symmetric central difference for the time derivative, leading to the explicit three-level scheme

$$\frac{U_j^{n+1} - U_j^{n-1}}{2\Delta t} = \frac{U_{j+1}^n - 2U_j^n + U_{j-1}^n}{(\Delta x)^2}. \tag{2.98}$$

It is easy to see that the truncation error of this approximation involves only even powers of both Δx and Δt, and hence has order $O\big((\Delta x)^2 + (\Delta t)^2\big)$. However, if we investigate the stability of the scheme we find a solution of the usual form (2.72) provided that

$$\frac{\lambda - 1/\lambda}{2\Delta t} = \frac{-4\sin^2\frac{1}{2}k\Delta x}{(\Delta x)^2} \tag{2.99}$$

or

$$\lambda^2 + 8\lambda\mu \sin^2\tfrac{1}{2}k\Delta x - 1 = 0. \tag{2.100}$$

This quadratic equation for λ has two roots, giving two solution modes for each value of k. The roots are both real with a negative sum, and their product is -1. Hence one of them has magnitude greater than unity, giving an unstable mode. The scheme is therefore useless in practice since it is always unstable, for every value of μ.

This result does not, of course, mean that every three-level explicit scheme is always unstable. We leave as an exercise the proof that the scheme

$$\frac{U_j^{n+1} - U_j^{n-1}}{2\Delta t} = \frac{\theta \delta_x^2 U_j^n + (1-\theta)\delta_x^2 U_j^{n-1}}{(\Delta x)^2} \qquad (2.101)$$

has both solution modes satisfying $|\lambda| \leq 1$ if $\theta \leq \frac{1}{2}$ and $4(1-\theta)\mu \leq 1$ (see also Exercise 6); but then this stability restriction is as bad as that for our first simple scheme.

2.13 More general boundary conditions

Let us now consider a more general model problem by introducing a derivative boundary condition at $x = 0$, of the form

$$\frac{\partial u}{\partial x} = \alpha(t)u + g(t), \quad \alpha(t) \geq 0. \qquad (2.102)$$

By using a forward space difference for the derivative, we can approximate this by

$$\frac{U_1^n - U_0^n}{\Delta x} = \alpha^n U_0^n + g^n \qquad (2.103)$$

and use this to give the boundary value U_0^n in the form

$$U_0^n = \beta^n U_1^n - \beta^n g^n \Delta x, \qquad (2.104a)$$

where

$$\beta^n = \frac{1}{1 + \alpha^n \Delta x}. \qquad (2.104b)$$

Then we can apply the θ-method (2.75) in just the same way as for Dirichlet boundary conditions. We need to solve the usual tridiagonal system of linear equations, but now this is a system of J equations in the J unknowns, namely the interior values at the new time step and the value at the left-hand boundary. Equation (2.104a) is then the first equation of the system, and it is clear that the augmented system is still tridiagonal and, because we have assumed $\alpha(t) \geq 0$ and therefore $0 < \beta^n \leq 1$, the coefficients still satisfy the conditions of (2.67) and (2.68), except that when $\alpha(t) = 0$ we have $b_0 = a_0$ and $e_0 = 1$.

To consider the accuracy and stability of the resulting scheme, we need to concentrate attention on the first interior point. We can use

(2.104a) to eliminate U_0^n; the second difference at the first interior point then has the form

$$\delta_x^2 U_1^n = U_2^n - (2 - \beta^n)U_1^n - \beta^n g^n \Delta x. \qquad (2.105)$$

Thus with the usual definition of truncation error, after some manipulation, the global error can be shown to satisfy, instead of (2.93), the new relation

$$\begin{aligned}
\left[1 + \theta\mu(2 - \beta^{n+1})\right] e_1^{n+1} &= \left[1 - (1 - \theta)\mu(2 - \beta^n)\right] e_1^n \\
&\quad + \theta\mu e_2^{n+1} + (1 - \theta)\mu e_2^n \\
&\quad - \Delta t T_1^{n+1/2}.
\end{aligned} \qquad (2.106)$$

This equation is different from that at other mesh points, which precludes our using Fourier analysis to analyse the system. But the maximum principle arguments of the preceding section can be used: we see first that if $\mu(1 - \theta) \leq \frac{1}{2}$ all the coefficients in (2.106) are nonnegative for any nonnegative value of α^n; and the sum of the coefficients on the right is no greater than that on the left if

$$\theta(1 - \beta^{n+1}) \geq -(1 - \theta)(1 - \beta^n), \qquad (2.107)$$

which again is always satisfied if $\alpha(t) \geq 0$. Hence we can deduce the bound (2.96) for the global error in terms of the truncation error as before. The importance of the assumption $\alpha(t) \geq 0$ is clear in these arguments: to assume otherwise would correspond to having heat inflow rather than outflow in proportion to surface temperature, which would lead to an exponentially increasing solution. This is very unlikely to occur in any real problem, and would in any case lead to a problem which is not well-posed.

It remains to estimate the truncation error $T_1^{n+1/2}$. Let us consider only the explicit case $\theta = 0$, for which we expand around the first interior point. Suppose we straightforwardly regard (2.103) as applying the boundary condition at $(0, t_n)$ and expand about this point for the exact solution to obtain

$$\frac{u_1^n - u_0^n}{\Delta x} - \alpha^n u_0^n - g^n = \left[\tfrac{1}{2}\Delta x\, u_{xx} + \tfrac{1}{6}(\Delta x)^2 u_{xxx} + \cdots\right]_0^n. \qquad (2.108)$$

Then we write the truncation error in the following form, in which an appropriate multiple of the approximation (2.103) to the boundary condition is added to the difference equation in order to cancel the terms in u_0^n,

$$T_1^{n+1/2} = \frac{u_1^{n+1} - u_1^n}{\Delta t} - \frac{\delta_x^2 u_1^n}{(\Delta x)^2} - \frac{\beta^n}{\Delta x}\left[\frac{u_1^n - u_0^n}{\Delta x} - \alpha^n u_0^n - g^n\right]$$

$$= \left[\tfrac{1}{2}\Delta t\, u_{tt} - \tfrac{1}{12}(\Delta x)^2 u_{xxxx} + \cdots\right]_1^n - \beta^n\left[\tfrac{1}{2}u_{xx} + \cdots\right]_0^n,$$

to obtain

$$T_1^{n+1/2} \approx -\tfrac{1}{2}\beta^n u_{xx}. \tag{2.109}$$

This does not tend to zero as the mesh size tends to zero and, although we could rescue our convergence proof by a more refined analysis, we shall not undertake this here.

However, a minor change can remedy the problem. We choose a new grid of points, which are still equally spaced, but with the boundary point $x = 0$ half-way between the first two grid points. The other boundary, $x = 1$, remains at the last grid point as before. We now replace the approximation to the boundary condition by the more accurate version

$$\frac{U_1^n - U_0^n}{\Delta x} = \tfrac{1}{2}\alpha^n\left(U_0^n + U_1^n\right) + g^n, \tag{2.110}$$

$$U_0^n = \frac{1 - \tfrac{1}{2}\alpha^n\Delta x}{1 + \tfrac{1}{2}\alpha^n\Delta x}U_1^n - \frac{\Delta x}{1 + \tfrac{1}{2}\alpha^n\Delta x}g^n. \tag{2.111}$$

Then (2.108) is replaced by an expansion about $j = \tfrac{1}{2}$, giving

$$\frac{u_1^n - u_0^n}{\Delta x} - \tfrac{1}{2}\alpha^n\left(u_0^n + u_1^n\right) - g^n = \left[\tfrac{1}{24}(\Delta x)^2 u_{xxx} - \tfrac{1}{8}\alpha^n(\Delta x)^2 u_{xx} + \cdots\right]_{1/2}^n \tag{2.112}$$

and hence

$$T^{n+1/2} = \left[\tfrac{1}{2}\Delta t u_{tt} - \tfrac{1}{12}(\Delta x)^2 u_{xxxx} + \cdots\right]_1^n$$
$$- \frac{1}{1 + \tfrac{1}{2}\alpha^n\Delta x}\left[\tfrac{1}{24}\Delta x\left(u_{xxx} - 3\alpha^n u_{xx}\right) + \cdots\right]_0^n$$
$$= O(\Delta x). \tag{2.113}$$

Only minor modifications are necessary to (2.106) and the proof of convergence is straightforward. Indeed, as we shall show in Chapter 6 where a sharper error analysis based on the maximum principle is presented, the error remains $O((\Delta x)^2)$ despite this $O(\Delta x)$ truncation error near the boundary.

An alternative, and more widely used, approach is to keep the first grid point at $x = 0$ but to introduce a fictitious value U_{-1}^n outside the domain so that we can use central differences to write

$$\frac{U_1^n - U_{-1}^n}{2\Delta x} = \alpha^n U_0^n + g^n. \tag{2.114}$$

Then the usual difference approximation is also applied at $x = 0$ so that U_{-1}^n can be eliminated. That is, for the θ-method we take

$$\frac{U_0^{n+1} - U_0^n}{\Delta t} - \frac{\delta_x^2}{(\Delta x)^2} \left[\theta U_0^{n+1} + (1 - \theta)U_0^n \right]$$
$$- \frac{2\theta}{\Delta x} \left[\frac{U_1^{n+1} - U_{-1}^{n+1}}{2\Delta x} - \alpha^{n+1}U_0^{n+1} - g^{n+1} \right]$$
$$- \frac{2(1 - \theta)}{\Delta x} \left[\frac{U_1^n - U_{-1}^n}{2\Delta x} - \alpha^n U_0^n - g^n \right] = 0. \tag{2.115}$$

Clearly for the truncation error we pick up terms like

$$\frac{2\theta}{\Delta x} \left[\frac{u_1^{n+1} - u_{-1}^{n+1}}{2\Delta x} - \alpha^{n+1}u_0^{n+1} - g^{n+1} \right] = \theta \left[\tfrac{1}{3}\Delta x u_{xxx} \right]_0^{n+1} + \cdots \tag{2.116}$$

to add to the usual truncation error terms. If we rewrite (2.115) in the form

$$\left[1 + 2\theta\mu \left(1 + \alpha^{n+1}\Delta x \right) \right] U_0^{n+1} = \left[1 - 2(1 - \theta)\mu \left(1 + \alpha^n \Delta x \right) \right] U_0^n$$
$$+ 2\theta\mu U_1^{n+1}$$
$$- 2\mu\Delta x \left[\theta g^{n+1} + (1 - \theta)g^n \right] \tag{2.117}$$

we also see that the error analysis based on a maximum principle still holds with only the slight strengthening of condition needed in Theorem 2.2 to

$$\mu(1 - \theta) \left(1 + \alpha^n \Delta x \right) \leq \tfrac{1}{2}. \tag{2.118}$$

In this section we have considered the solution of the heat equation with a derivative boundary condition at the left-hand end, and a Dirichlet condition at the other. The same idea can be applied to a problem with a derivative condition at the right-hand end, or at both ends.

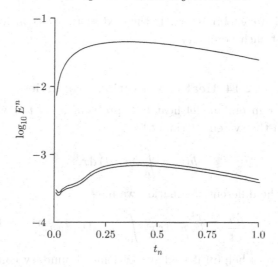

Fig. 2.10. The effect of a Neumann boundary condition approximation on the error for the Crank–Nicolson scheme with $J = 10$, $\Delta x = 0.1$; the top curve is for (2.103) and the lower two for (2.114) and (2.110).

On carrying through the similar analysis, we soon discover why the condition at $x = 1$ must be of the form

$$\frac{\partial u}{\partial x} = \beta(t)u + g(t), \quad \beta(t) \leq 0. \tag{2.119}$$

As an illustration of different methods of treating the boundary condition, we compute solutions to the problem $u_t = u_{xx}$ on $0 < x < 1$, with initial condition $u(x,0) = 1 - x^2$, and boundary conditions $u_x(0,t) = 0$, $u(1,t) = 0$, giving a Neumann condition at the left-hand end, and a Dirichlet condition at the right-hand end. We use the Crank–Nicolson method, with $J = 10$ and $\mu = 1$, so that $\Delta t = 0.01$. The maximum error in the numerical solution, as a function of t_n, is shown in Fig. 2.10 for the three methods described above: namely, the use of the forward difference approximation to $u_x(0,t)$, the use of the central difference incorporating the fictitious value U_{-1}^n, and the placing of the boundary half-way between the first two mesh points. The numerical results from the second and third of these methods are very similar, but show a quite dramatic difference from those for the first method; the error in this case is some 50 times larger.

Notice also that the maximum error in each method increases with n for part of the range, before beginning to decrease again, rather slowly; compare the behaviour with that in Fig. 2.4 where Dirichlet conditions were applied at each boundary. In the next section we consider Neumann conditions at each boundary.

2.14 Heat conservation properties

Suppose that in our model heat flow problem $u_t = u_{xx}$ we define the total heat in the system at time t by

$$h(t) = \int_0^1 u(x,t)\,\mathrm{d}x. \tag{2.120}$$

Then from the differential equation we have

$$\frac{\mathrm{d}h}{\mathrm{d}t} = \int_0^1 u_t\,\mathrm{d}x = \int_0^1 u_{xx}\,\mathrm{d}x = \left[u_x\right]_0^1. \tag{2.121}$$

This is not very helpful if we have Dirichlet boundary conditions: but suppose we are given Neumann boundary conditions at each end; say, $u_x(0,t) = g_0(t)$ and $u_x(1,t) = g_1(t)$. Then we have

$$\frac{\mathrm{d}h}{\mathrm{d}t} = g_1(t) - g_0(t), \tag{2.122}$$

so that h is given by integrating an ordinary differential equation.

Now suppose we carry out a similar manipulation for the θ-method equations (2.75), introducing the total heat by means of a summation over the points for which (2.75) holds:

$$H^n = \sum_1^{J-1} \Delta x U_j^n. \tag{2.123}$$

Then, recalling from the definitions of the finite difference notation that $\delta_x^2 U_j = \Delta_{+x} U_j - \Delta_{+x} U_{j-1}$, we have

$$\begin{aligned}
H^{n+1} - H^n &= \frac{\Delta t}{\Delta x} \sum_1^{J-1} \delta_x^2 \left[\theta U_j^{n+1} + (1-\theta)U_j^n\right] \\
&= \frac{\Delta t}{\Delta x} \left\{ \Delta_{+x} \left[\theta U_{J-1}^{n+1} + (1-\theta)U_{J-1}^n\right] \right. \\
&\quad \left. - \Delta_{+x} \left[\theta U_0^{n+1} + (1-\theta)U_0^n\right] \right\}.
\end{aligned} \tag{2.124}$$

The rest of the analysis will depend on how the boundary condition is approximated. Consider the simplest case as in (2.103): namely we set $U_1^n - U_0^n = \Delta x\, g_0^n$, $U_J^n - U_{J-1}^n = \Delta x\, g_1^n$. Then we obtain

$$H^{n+1} - H^n = \Delta t \left[\theta \left(g_1^{n+1} - g_0^{n+1}\right) + (1-\theta)\left(g_1^n - g_0^n\right)\right] \tag{2.125}$$

as an approximation to (2.122); this approximation may be very accurate, even though we have seen that U^n may not give a good pointwise approximation to u^n. In particular, if g_0 and g_1 are independent of t the change in H in one time step exactly equals that in $h(t)$. How should we interpret this?

Clearly to make the most of this matching we should relate (2.123) as closely as possible to (2.120). If u and U were constants that would suggest we take $(J-1)\Delta x = 1$, rather than $J\Delta x = 1$ as we have been assuming; and we should compare U^n_j with

$$u^n_j := \frac{1}{\Delta x} \int_{(j-1)\Delta x}^{j\Delta x} u(x, t_n)\, dx, \quad j = 1, 2, \ldots, J-1, \qquad (2.126)$$

so that it is centred at $(j - \frac{1}{2})\Delta x$ and we have

$$h(t_n) = \sum_{1}^{J-1} \Delta x\, u^n_j. \qquad (2.127)$$

Note that this interpretation matches very closely the scheme that we were led to in (2.110)–(2.113) by analysing the truncation error. It would also mean that for initial condition we should take $U^0_j = u^0_j$ as defined by (2.126). Then for time-independent boundary conditions we have $H^n = h(t_n)$ for all n. Moreover, it is easy to see that the function

$$\hat{u}(x, t) := (g_1 - g_0)\, t + \tfrac{1}{2}\, (g_1 - g_0)\, x^2 + g_0 x + C \qquad (2.128)$$

with any constant C satisfies the differential equation, and the two boundary conditions. It can also be shown that the exact solution of our problem, with any given initial condition, will tend to such a solution as $t \to \infty$. Since the function (2.128) is linear in t and quadratic in x it will also satisfy the finite difference equations exactly; hence the error in a numerical solution produced and interpreted in this way will decrease to zero as t increases. As we have seen above, the usual finite difference approximations, with Neumann boundary conditions and the usual interpretation of the errors, may be expected to give errors which initially increase with n, and then damp only very slowly.

These observations are illustrated by a solution of the heat equation with homogeneous Neumann boundary conditions $u_x = 0$ at $x = 0$ and $x = 1$, and with initial value $u(x, 0) = 1 - x^2$. Fig. 2.11 shows the maximum error as a function of t_n for three cases, each using $J = 10$ for the Crank–Nicolson method with $\mu = \frac{1}{2}$. The top curve corresponds to using the Neumann conditions (2.103) as above and interpreting the

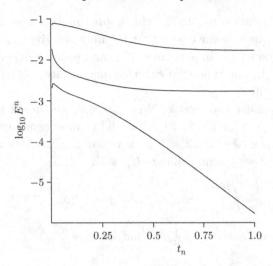

Fig. 2.11. Effect of error interpretation for a pure Neumann problem: the top curve corresponds to boundary conditions (2.103) and the second to (2.114), both with the usual initial data and definition of E^n; the bottom curve is computed as for the top curve but with initial data from (2.126) and the error reinterpreted through the heat conservation principle.

error E^n in the usual way, with $\Delta x = 0.1$, $\Delta t = 0.005$. The second curve is the same apart from using the approximaton (2.114) for the boundary conditions. Clearly both give a substantial residual error. However, the bottom curve corresponds to the same method as the top but with the initial data obtained from the u_j^n of (2.126), and the error reinterpreted to reflect the heat conservation principle as described above: namely, $E^n := \max\{|U_j^n - u_j^n|, \quad j = 0, 1, 2, \ldots, J\}$; and $t_n = n\Delta t = n\mu(\Delta x)^2 = \frac{1}{2}n(J-1)^{-2}$. We see clearly the decay of the error as predicted by the argument given above.

2.15 More general linear problems

The form of the heat equation which we have considered so far corresponds to a physical situation where all the physical properties of the material are constant in time, and independent of x. More generally these properties may be functions of x, or t, or both. In particular, a dependence on x is often used to model the nearly one-dimensional flow of heat in a thin bar whose cross-sectional area depends on x. We shall therefore examine briefly how the methods so far discussed may be adapted to problems of increasing generality.

First of all, consider the problem

$$\frac{\partial u}{\partial t} = b(x,t)\frac{\partial^2 u}{\partial x^2}, \tag{2.129}$$

where the function $b(x,t)$ is, as usual, assumed to be strictly positive. Then the explicit scheme (2.19) is extended in an obvious way to give

$$U_j^{n+1} = U_j^n + \frac{\Delta t}{(\Delta x)^2}b_j^n(U_{j+1}^n - 2U_j^n + U_{j-1}^n), \tag{2.130}$$

where $b_j^n = b(x_j, t_n)$. The practical implementation of this scheme is just as easy as before, and the analysis of the error is hardly altered. The same expansion in Taylor series leads, as for (2.19), to the expression

$$T(x,t) = \tfrac{1}{2}\Delta t\, u_{tt} - \tfrac{1}{12}b(x,t)(\Delta x)^2 u_{xxxx} + \cdots. \tag{2.131}$$

The analysis leading to (2.44) still applies, but the stability condition has to be replaced by

$$\frac{\Delta t}{(\Delta x)^2}b(x,t) \le \tfrac{1}{2} \tag{2.132}$$

for all values of x and t in the region. The final error bound becomes

$$E^n \le \tfrac{1}{2}\Delta t\left[M_{tt} + \frac{B(\Delta x)^2}{6\Delta t}\,M_{xxxx}\right]t_F \tag{2.133}$$

where B is a uniform upper bound for $b(x,t)$ in the region $[0,1]\times[0,t_F]$.

The θ-method can be applied to this more general problem in several slightly different ways. Evidently equation (2.75) can be generalised to

$$U_j^{n+1} - U_j^n = \frac{\Delta t}{(\Delta x)^2}b^*\left[\theta\delta_x^2 U_j^{n+1} + (1-\theta)\delta_x^2 U_j^n\right], \tag{2.134}$$

but it is not obvious what is the best value to use for b^*. In our previous analysis of the truncation error of this scheme we expanded in Taylor series about the centre point $(x_j, t_{n+1/2})$. This suggests the choice

$$b^* = b_j^{n+1/2}; \tag{2.135}$$

and in fact it is easy to see that with this choice our former expansion of the truncation error is unaltered, except for the inclusion of the extra factor b in (2.84), which becomes

$$\begin{aligned}
T_j^{n+1/2} = [(\tfrac{1}{2} - \theta)\Delta t\, u_{xxt} &- \tfrac{b}{12}(\Delta x)^2 u_{xxxx} + \tfrac{1}{24}(\Delta t)^2 u_{ttt} \\
&- \tfrac{b}{8}(\Delta t)^2 u_{xxtt} + \tfrac{1}{12}(\tfrac{1}{2} - \theta)\Delta t(\Delta x)^2 u_{xxxxt} \\
&- \tfrac{2b}{6!}(\Delta x)^4 u_{xxxxxx} + \cdots]_j^{n+1/2}.
\end{aligned} \tag{2.136}$$

The proof of convergence by means of a maximum principle is also unaltered, except that the stability condition now requires that

$$\frac{\Delta t}{(\Delta x)^2}(1-\theta)b(x,t) \leq \tfrac{1}{2} \tag{2.137}$$

for all points (x,t) in the region considered.

This choice of b^* requires the computation of $b(x,t)$ for values of t half-way between time steps. This may be awkward in some problems, and an obvious alternative is to use

$$b^* = \tfrac{1}{2}(b_j^{n+1} + b_j^n). \tag{2.138}$$

Now we need another Taylor expansion, about the centre point, giving

$$b^* = \left[b + \tfrac{1}{4}(\Delta t)^2 b_{tt} + \cdots\right]_j^{n+1/2} \tag{2.139}$$

which will lead to an additional higher order term, involving b_{tt}, appearing in the expansion of the truncation error.

The most general form of the linear parabolic equation is

$$\frac{\partial u}{\partial t} = b(x,t)\frac{\partial^2 u}{\partial x^2} - a(x,t)\frac{\partial u}{\partial x} + c(x,t)u + d(x,t), \tag{2.140}$$

where as before $b(x,t)$ is assumed to be always positive. The notation used here is chosen to match that used in later chapters, and specifically in (5.48) of Section 5.7. In particular the negative sign in front of $a(x,t)$ is convenient but unimportant, since $a(x,t)$ may take either sign; only $b(x,t)$ is required to be positive. We can easily construct an explicit scheme for this equation; only the term in $\partial u/\partial x$ needs any new consideration. As we have used the central difference approximation for the second derivative, it is natural to use the central difference approximation for the first derivative, leading to the scheme

$$\frac{U_j^{n+1} - U_j^n}{\Delta t} = \frac{b_j^n}{(\Delta x)^2}(U_{j+1}^n - 2U_j^n + U_{j-1}^n)$$

$$- \frac{a_j^n}{2\Delta x}(U_{j+1}^n - U_{j-1}^n) + c_j^n U_j^n + d_j^n. \tag{2.141}$$

The calculation of the leading terms in the truncation error is straightforward, and is left as an exercise. However, a new difficulty arises in the analysis of the behaviour of the error e_j^n. Just as in the analysis of the simpler problem, which led to (2.42), we find that

$$
\begin{aligned}
e_j^{n+1} &= e_j^n + \mu_j^n\,(e_{j+1}^n - 2e_j^n + e_{j-1}^n) - \tfrac{1}{2}\nu_j^n(e_{j+1}^n - e_{j-1}^n) \\
&\quad + \Delta t c_j^n e_j^n - \Delta t T_j^n \\
&= (1 - 2\mu_j^n + \Delta t c_j^n)e_j^n \\
&\quad + (\mu_j^n - \tfrac{1}{2}\nu_j^n)e_{j+1}^n + (\mu_j^n + \tfrac{1}{2}\nu_j^n)e_{j-1}^n - \Delta t T_j^n, \qquad (2.142)
\end{aligned}
$$

where we have written

$$
\mu_j^n = \frac{\Delta t}{(\Delta x)^2}b_j^n, \qquad \nu_j^n = \frac{\Delta t}{\Delta x}a_j^n. \qquad (2.143)
$$

In order to go on to obtain similar bounds for e_j^n as before, we need to ensure that the coefficients of the three terms in e^n on the right of this equation are all nonnegative and have a sum no greater than unity. We always assume that the function $b(x,t)$ is strictly positive, but we cannot in general assume anything about the sign of $a(x,t)$. We are therefore led to the conditions:

$$
\tfrac{1}{2}\left|\nu_j^n\right| \le \mu_j^n, \qquad (2.144)
$$

$$
2\mu_j^n - \Delta t c_j^n \le 1, \qquad (2.145)
$$

as well as $c_j^n \le 0$. The second of these conditions is only slightly more restrictive than in the simpler case, because of the condition $c_j^n \le 0$; indeed, if we had $0 \le c(x,t) \le C$ condition (2.145) would represent a slight relaxation of the condition on μ, but then one can only establish $E^{n+1} \le (1 + C\Delta t)E^n + T\Delta t$. However, the first condition is much more serious. If we replace ν and μ by their expressions in terms of Δt and Δx this becomes

$$
\Delta x \le \frac{2b_j^n}{|a_j^n|}, \quad \text{or} \quad \frac{|a_j^n|\,\Delta x}{b_j^n} \le 2, \qquad (2.146)
$$

and this condition must hold for all values of n and j. We therefore have a restriction on the size of Δx, which also implies a restriction on the size of Δt.

In many practical problems the function $b(x,t)$ may be very small compared with $a(x,t)$. This will happen, for example, in the flow of most fluids, which have a very small viscosity. In such situations a key dimensionless parameter is the *Péclet number* UL/ν, where U is a velocity, L a length scale and ν the viscosity. These are close to what are known as *singular perturbation problems*, and cannot easily be solved by this explicit, central difference method: for (2.146) imposes a limit of 2 on a *mesh Péclet number* in which the length scale is the mesh length. Suppose, for example, that $b = 0.001, a = 1, c = 0$. Then our conditions require that $\Delta x \leq 0.002$, and therefore that $\Delta t \leq 0.000002$. We thus need at least 500 mesh points in the x-direction, and an enormous number of time steps to reach any sensible time t_F.

A simple way of avoiding this problem is to use forward or backward differences for the first derivative term, instead of the central difference. Suppose, for example, it is known that $a(x,t) \geq 0$ and $c(x,t) = 0$. We then use the backward difference, and our difference formula becomes

$$\frac{U_j^{n+1} - U_j^n}{\Delta t} = \frac{b_j^n}{(\Delta x)^2}(U_{j+1}^n - 2U_j^n + U_{j-1}^n)$$

$$- \frac{a_j^n}{\Delta x}(U_j^n - U_{j-1}^n) + c_j^n U_j^n + d_j^n, \qquad (2.147)$$

which leads to

$$e_j^{n+1} = e_j^n + \mu_j^n(e_{j+1}^n - 2e_j^n + e_{j-1}^n) - \nu_j^n(e_j^n - e_{j-1}^n) - \Delta t T_j^n$$
$$= (1 - 2\mu_j^n - \nu_j^n)e_j^n$$
$$+ \mu_j^n e_{j+1}^n + (\mu_j^n + \nu_j^n)e_{j-1}^n - \Delta t T_j^n. \qquad (2.148)$$

In order to ensure that all the coefficients on the right of this equation are nonnegative, we now need only

$$2\mu_j^n + \nu_j^n \leq 1. \qquad (2.149)$$

This requires a more severe restriction on the size of the time step when $a \neq 0$, but no restriction on the size of Δx.

If the function $a(x,t)$ changes sign, we can use the backward difference where a is positive, and the forward difference where it is negative; this idea is known as *upwind differencing* . Unfortunately we have to pay a price for this lifting of the restriction needed to ensure a maximum principle. The truncation error is now of lower order: the forward difference introduces an error of order Δx, instead of the order $(\Delta x)^2$ given by the central difference. However, we shall discuss this issue in the chapter on hyperbolic equations.

A general parabolic equation may also often appear in the self-adjoint form

$$\frac{\partial u}{\partial t} = \frac{\partial}{\partial x}\left(p(x,t)\frac{\partial u}{\partial x}\right) \qquad (2.150)$$

where, as usual, we assume that the function $p(x,t)$ is strictly positive. It is possible to write this equation in the form just considered, as

$$\frac{\partial u}{\partial t} = p\frac{\partial^2 u}{\partial x^2} + \frac{\partial p}{\partial x}\frac{\partial u}{\partial x}, \qquad (2.151)$$

but it is usually better to construct a difference approximation to the equation in its original form. We can write

$$\left[p\frac{\partial u}{\partial x}\right]_{j+1/2}^n \approx p_{j+1/2}^n \left(\frac{u_{j+1}^n - u_j^n}{\Delta x}\right), \qquad (2.152)$$

and a similar approximation with j replaced by $j-1$ throughout. If we subtract these two, and divide by Δx, we obtain an approximation to the right-hand side of the equation, giving the explicit difference scheme

$$\frac{U_j^{n+1} - U_j^n}{\Delta t} = \frac{1}{(\Delta x)^2}\left[p_{j+1/2}^n(U_{j+1}^n - U_j^n) - p_{j-1/2}^n(U_j^n - U_{j-1}^n)\right]. \qquad (2.153)$$

We will write

$$\mu' = \frac{\Delta t}{(\Delta x)^2}$$

which gives in explicit form

$$U_j^{n+1} = \left(1 - \mu'(p_{j+1/2}^n + p_{j-1/2}^n)\right)U_j^n + \mu' p_{j+1/2}^n U_{j+1}^n + \mu' p_{j-1/2}^n U_{j-1}^n. \qquad (2.154)$$

This shows that the form of error analysis which we have used before will again apply here, with each of the coefficients on the right-hand side being nonnegative provided that

$$\mu'P \le \tfrac{1}{2}, \qquad (2.155)$$

where P is an upper bound for the function $p(x,t)$ in the region. So this scheme gives just the sort of time step restriction which we should expect, without any restriction on the size of Δx.

The same type of difference approximation can be applied to give an obvious generalisation of the θ-method. The details are left as an exercise, as is the calculation of the leading terms of the truncation error (see Exercises 7 and 8).

2.16 Polar co-ordinates

One-dimensional problems often result from physical systems in three dimensions which have cylindrical or spherical symmetry. In polar co-ordinates the simple heat equation becomes

$$\frac{\partial u}{\partial t} = \frac{1}{r^\alpha} \frac{\partial}{\partial r} \left(r^\alpha \frac{\partial u}{\partial r} \right) \qquad (2.156)$$

or

$$u_t = u_{rr} + \frac{\alpha}{r} u_r, \qquad (2.157)$$

where $\alpha = 0$ reduces to the case of plane symmetry which we have considered so far, while $\alpha = 1$ corresponds to cylindrical symmetry and $\alpha = 2$ to spherical symmetry. The methods just described can easily be applied to this equation, either in the form (2.156), or in the form (2.157). Examination of the stability restrictions in the two cases shows that there is not much to choose between them in this particular situation. However, in each case there is clearly a problem at the origin $r = 0$.

A consideration of the symmetry of the solution, in either two or three dimensions, shows that $\partial u/\partial r = 0$ at the origin; alternatively, (2.157) shows that either u_{rr} or u_t, or both, would be infinite at $r = 0$, were u_r non-zero. Now keep t constant, treating u as a function of r only, and expand in a Taylor series around $r = 0$, giving

$$u(r) = u(0) + r u_r(0) + \tfrac{1}{2} r^2 u_{rr}(0) + \cdots$$
$$= u(0) + \tfrac{1}{2} r^2 u_{rr}(0) + \cdots \qquad (2.158)$$

and

$$\frac{1}{r^\alpha} \frac{\partial}{\partial r} \left(r^\alpha \frac{\partial u}{\partial r} \right) = \frac{1}{r^\alpha} \frac{\partial}{\partial r} \left[r^\alpha u_r(0) + r^{\alpha+1} u_{rr}(0) + \cdots \right]$$
$$= \frac{1}{r^\alpha} \left[(\alpha + 1) r^\alpha u_{rr}(0) + \cdots \right]$$
$$= (\alpha + 1) u_{rr}(0) + \cdots . \qquad (2.159)$$

Writing Δr for r in (2.158) we get

$$u(\Delta r) - u(0) = \tfrac{1}{2} (\Delta r)^2 u_{rr}(0) + \cdots \qquad (2.160)$$

and we thus obtain a difference approximation to be used at the left end of the domain,

$$\frac{U_0^{n+1} - U_0^n}{\Delta t} = \frac{2(\alpha + 1)}{(\Delta r)^2} (U_1^n - U_0^n). \qquad (2.161)$$

This would also allow any of the θ-methods to be applied.

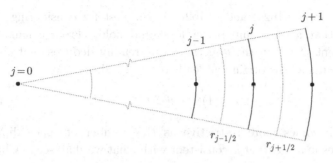

Fig. 2.12. Polar co-ordinates.

An alternative, more physical, viewpoint springs directly from the form (2.156). Consider the heat balance for an annular region between two surfaces at $r = r_{j-1/2}$ and $r = r_{j+1/2}$ as in Fig. 2.12: the term $r^\alpha \partial u / \partial r$ on the right-hand side of (2.156) is proportional to a heat flux times a surface area; and the difference between the fluxes at surfaces with radii $r_{j-1/2}$ and $r_{j+1/2}$ is applied to raising the temperature in a volume which is proportional to $(r_{j+1/2}^{\alpha+1} - r_{j-1/2}^{\alpha+1})/(\alpha + 1)$. Thus on a uniform mesh of spacing Δr, a direct differencing of the right-hand side of (2.156) gives

$$
\frac{\partial U_j}{\partial t} \approx \frac{\alpha + 1}{r_{j+1/2}^{\alpha+1} - r_{j-1/2}^{\alpha+1}} \delta_r \left(r_j^\alpha \frac{\delta_r U_j}{\Delta r} \right)
$$

$$
= \frac{(\alpha + 1) \left[r_{j+1/2}^\alpha U_{j+1} - \left(r_{j+1/2}^\alpha + r_{j-1/2}^\alpha \right) U_j + r_{j-1/2}^\alpha U_{j-1} \right]}{\left[r_{j+1/2}^\alpha + r_{j+1/2}^{\alpha-1} r_{j-1/2} + \cdots + r_{j-1/2}^\alpha \right] (\Delta r)^2}
$$

$$
\text{for } j = 1, 2, \ldots . \qquad (2.162a)
$$

At the origin where there is only one surface (a cylinder of radius $r_{1/2} = \frac{1}{2}\Delta r$ when $\alpha = 1$, a sphere of radius $r_{1/2}$ when $\alpha = 2$) one has immediately

$$
\frac{\partial U_0}{\partial t} \approx \frac{\alpha + 1}{r_{1/2}^{\alpha+1}} r_{1/2}^\alpha \frac{U_1 - U_0}{\Delta r} = 2(\alpha + 1) \frac{U_1 - U_0}{(\Delta r)^2}, \qquad (2.162b)
$$

which is in agreement with (2.161). Note also that (2.162a) is identical with difference schemes obtained from either (2.156) or (2.157) in the case of cylindrical symmetry ($\alpha = 1$); but there is a difference in the spherical case because $r_{j+1/2}^2 + r_{j+1/2} r_{j-1/2} + r_{j-1/2}^2$ is not the same as $3r_j^2$.

The form (2.162a) and (2.162b) is simplest for considering the condition that a maximum principle should hold. From calculating the coefficient of U_j^n in the θ-method, one readily deduces that the worst case occurs at the origin and leads to the condition

$$2(\alpha + 1)(1 - \theta)\Delta t \le (\Delta r)^2. \qquad (2.162c)$$

This becomes more restrictive as the number of space dimensions increases in a way that is consistent with what we shall see in Chapter 3.

2.17 Nonlinear problems

In the general linear equation we considered in Section 2.15 the physical properties depended on x and t. It is also very common for these properties to depend on the unknown function $u(x, t)$. This leads to the consideration of nonlinear problems.

We shall just consider one example, the equation

$$u_t = b(u)u_{xx} \qquad (2.163)$$

where the coefficient $b(u)$ depends on the solution u only and must be assumed strictly positive for all u. This simplification is really only for ease of notation; it is not much more difficult to treat the case in which b is a function of x and t as well as of u.

The explicit method is little affected; it becomes, in the same notation as before,

$$U_j^{n+1} = U_j^n + \mu' b(U_j^n)\left(U_{j+1}^n - 2U_j^n + U_{j-1}^n\right). \qquad (2.164)$$

The actual calculation is no more difficult than before, the only extra work being the computation of the function $b(U_j^n)$. The truncation error also has exactly the same form as before and the conditions for the values $\{U_j^n\}$ to satisfy a maximum principle are unchanged. However, the analysis of the behaviour of the global error e_j^n is more difficult, as it propagates in a nonlinear way as n increases.

Writing u_j^n for the value of the exact solution $u(x_j, t_n)$ we know that U_j^n and u_j^n satisfy the respective equations

$$U_j^{n+1} = U_j^n + \mu' b(U_j^n)(U_{j+1}^n - 2U_j^n + U_{j-1}^n), \qquad (2.165)$$
$$u_j^{n+1} = u_j^n + \mu' b(u_j^n)(u_{j+1}^n - 2u_j^n + u_{j-1}^n) + \Delta t\, T_j^n, \qquad (2.166)$$

where T_j^n is the truncation error. But we cannot simply subtract these equations to obtain a relation for e_j^n, since the two coefficients $b(\cdot)$ are different. However we can first write

$$b(u_j^n) = b(U_j^n) + (u_j^n - U_j^n)\frac{\partial b}{\partial u}(\eta) \qquad (2.167)$$

$$= b(U_j^n) - e_j^n q_j^n \qquad (2.168)$$

where

$$q_j^n = \frac{\partial b}{\partial u}(\eta) \qquad (2.169)$$

and η is some number between U_j^n and u_j^n.

We can now subtract (2.166) from (2.165), and obtain

$$e_j^{n+1} = e_j^n + \mu' b(U_j^n)(e_{j+1}^n - 2e_j^n + e_{j-1}^n)$$
$$+ \mu' e_j^n q_j^n (u_{j+1}^n - 2u_j^n + u_{j-1}^n) - \Delta t\, T_j^n. \qquad (2.170)$$

The coefficients of e_{j-1}^n, e_j^n, e_{j+1}^n arising from the first two terms on the right are now nonnegative provided that

$$\Delta t\, [\max b(U_j^n)] \le \tfrac{1}{2}(\Delta x)^2. \qquad (2.171)$$

This is our new stability condition, and the condition for the approximation to satisfy a maximum principle; in general it will need to be checked (and Δt adjusted) at each time step. However, assuming that we can use a constant step Δt which satisfies (2.171) for all j and n, and that we have bounds

$$\left|u_{j+1} - 2u_j^n + u_{j-1}^n\right| \le M_{xx}(\Delta x)^2, \quad \left|q_j^n\right| \le K, \qquad (2.172)$$

we can write

$$E^{n+1} \le [1 + K M_{xx}\Delta t]\, E^n + \Delta t\, T \qquad (2.173)$$

in our previous notation. Moreover,

$$(1 + K M_{xx}\Delta t)^n \le e^{K M_{xx} n\Delta t} \le e^{K M_{xx} t_F} \qquad (2.174)$$

and this allows a global error bound to be obtained in terms of T.

However, although the stability condition (2.171) is not much stronger than that for the linear problem, the error bound is much worse unless the a priori bounds on $|\partial b/\partial u|$ and $|u_{xx}|$ are very small. Furthermore, our example (2.163) is rather special; equally common would be the case $u_t = (b(u)u_x)_x$, and that gives an extra term $(\partial b/\partial u)(u_x)^2$ which can make very great changes to the problem and its analysis.

To summarise, then, the actual application of our explicit scheme to nonlinear problems gives little difficulty. Indeed the main practical use of numerical methods for partial differential equations is for nonlinear problems, where alternative methods break down. Even our implicit methods are not very much more difficult to use; there is just a system of nonlinear equations to solve with a good first approximation given from the previous time level. However, the analysis of the convergence and stability behaviour of these schemes is very much more difficult than for the linear case.

Bibliographic notes and recommended reading

The general reference for all the material on initial value problems in this book is the classic text by Richtmyer and Morton (1967). Another classic text covering the whole range of partial differential equation problems is that by Collatz (1966). In both of them many more difference schemes to approximate the problems treated in this chapter will be found.

For a wide-ranging exposition of diffusion problems and applications in which they arise, the reader is referred to the book by Crank (1975), where more discussion on the use of the Crank–Nicolson scheme may be found.

The earliest reference that we have to the important Thomas algorithm is to a report from Columbia University, New York in 1949; but since it corresponds to direct Gaussian elimination without pivoting, many researchers were undoubtedly aware of it at about this time. For a more general discussion of Gaussian elimination for banded matrices the reader is referred to standard texts on numerical analysis, several of which are listed in the Bibliography at the end of the book, or to more specialised texts on matrix computations such as that by Golub and Van Loan (1996).

A fuller discussion of nonlinear problems is given in the book by Ames (1992), where reference is made to the many examples of physical problems which are modelled by nonlinear parabolic equations contained in Ames (1965) and Ames (1972).

Exercises

2.1   (i) The function $u^0(x)$ is defined on [0,1] by

$$u^0(x) = \begin{cases} 2x & \text{if } 0 \le x \le \frac{1}{2}, \\ 2 - 2x & \text{if } \frac{1}{2} \le x \le 1. \end{cases}$$

Show that

$$u^0(x) = \sum_{m=1}^{\infty} a_m \sin m\pi x$$

where $a_m = (8/m^2\pi^2) \sin \frac{1}{2}m\pi$.

(ii) Show that

$$\int_{2p}^{2p+2} \frac{1}{x^2} dx > \frac{2}{(2p+1)^2}$$

and hence that

$$\sum_{p=p_0}^{\infty} \frac{1}{(2p+1)^2} < \frac{1}{4p_0}.$$

(iii) Deduce that $u^0(x)$ is approximated on the interval $[0,1]$ to within 0.001 by the sine series in part (i) truncated after $m = 405$.

2.2   (i) Show that for every positive value of $\mu = \Delta t/(\Delta x)^2$ there exists a constant $C(\mu)$ such that, for all positive values of k and Δx,

$$\left| 1 - 4\mu \sin^2 \tfrac{1}{2}k\Delta x - e^{-k^2\Delta t} \right| \leq C(\mu)k^4(\Delta t)^2.$$

Verify that when $\mu = \frac{1}{4}$ this inequality is satisfied by $C = \frac{1}{2}$.

(ii) The explicit central difference method is used to construct a solution of the equation $u_t = u_{xx}$ on the region $0 \leq x \leq 1$, $t \geq 0$. The boundary conditions specify that $u(0,t) = u(1,t) = 0$, and $u(x,0) = u^0(x)$, the same function as given in Exercise 1. Take $\epsilon = 0.01$, and show that in the sine series given there

$$\sum_{m=2p_0+1}^{\infty} |a_m| \leq \frac{\epsilon}{4} \text{ if } p_0 = 82,$$

and that then

$$\sum_{m=1}^{2p_0-1} |a_m|m^4 \leq 8p_0(2p_0 + 1)(2p_0 - 1)/3\pi^2.$$

Deduce that over the range $0 \leq t \leq 1$ the numerical solution will have an error less than 0.01 when $\mu = \frac{1}{4}$ provided that $\Delta t \leq 1.7 \times 10^{-10}$.

(iii) Verify by calculation that the numerical solution has error less than 0.01 over this range when $\mu = \frac{1}{4}$ and $\Delta t = 0.0025$.

[Observe that for this model problem the largest error is always in the first time step.]

2.3   Suppose that the mesh points x_j are chosen to satisfy

$$0 = x_0 < x_1 < x_2 < \cdots < x_{J-1} < x_J = 1$$

but are otherwise arbitrary. The equation $u_t = u_{xx}$ is approximated over the interval $0 \le t \le t_F$ by

$$\frac{U_j^{n+1} - U_j^n}{\Delta t} = \frac{2}{\Delta x_{j-1} + \Delta x_j} \left(\frac{U_{j+1}^n - U_j^n}{\Delta x_j} - \frac{U_j^n - U_{j-1}^n}{\Delta x_{j-1}} \right)$$

where $\Delta x_j = x_{j+1} - x_j$. Show that the leading terms of the truncation error of this approximation are

$$T_j^n = \tfrac{1}{2}\Delta t\, u_{tt} - \tfrac{1}{3}(\Delta x_j - \Delta x_{j-1})u_{xxx}$$
$$- \tfrac{1}{12}[(\Delta x_j)^2 + (\Delta x_{j-1})^2 - \Delta x_j \Delta x_{j-1}]u_{xxxx}.$$

Now suppose that the boundary conditions prescribe the values of $u(0,t)$, $u(1,t)$ and $u(x,0)$. Write $\Delta x = \max \Delta x_j$, and suppose that the mesh is sufficiently smooth so that $|\Delta x_j - \Delta x_{j-1}| \le \alpha(\Delta x)^2$, for $j = 1, 2, \ldots, J-1$, where α is a constant. Show that

$$|U_j^n - u(x_j, t_n)| \le \left(\tfrac{1}{2}\Delta t\, M_{tt} \right.$$
$$\left. + (\Delta x)^2 \left\{ \tfrac{1}{3}\alpha M_{xxx} + \tfrac{1}{12}[1 + \alpha \Delta x]M_{xxxx} \right\} \right) t_F$$

in the usual notation, provided that the stability condition

$$\Delta t \le \tfrac{1}{2}\Delta x_{j-1}\Delta x_j, \quad j = 1, 2, \ldots, J-1,$$

is satisfied.

2.4   The numbers a_j, b_j, c_j satisfy

$$a_j > 0, \; c_j > 0, \; b_j > a_j + c_j, \quad j = 1, 2, \ldots, J-1,$$

and

$$e_j = \frac{c_j}{b_j - a_j e_{j-1}}, \quad j = 1, 2, \ldots, J-1,$$

with $e_0 = 0$. Show by induction that $0 < e_j < 1$ for $j = 1, 2, \ldots, J-1$.

Show, further, that the conditions

$$b_j > 0, \; b_j \ge |a_j| + |c_j|, \quad j = 1, 2, \ldots, J-1,$$

are sufficient for $|e_0| \leq 1$ to imply that $|e_j| \leq 1$ for $j = 1, 2, \ldots, J - 1$.

2.5 Consider the equation $u_t = u_{xx}$, with the boundary condition $u(1, t) = 0$ for all $t \geq 0$, and

$$\frac{\partial u}{\partial x} = \alpha(t)u + g(t) \quad \text{at } x = 0, \text{ for all } t \geq 0,$$

with $\alpha(t) \geq 0$. Show in detail how the Thomas algorithm is used when solving the equation by the θ-method. In particular, derive the starting conditions which replace equation (2.71).

2.6 (i) By considering separately the cases of real roots and complex roots, or otherwise, show that both the roots of the quadratic equation $z^2 + bz + c = 0$ with real coefficients lie in or on the unit circle if and only if $|c| \leq 1$ and $|b| \leq 1 + c$.
(ii) Show that the scheme

$$U_j^{n+1} - U_j^{n-1} = \tfrac{1}{3}\mu \left\{ \delta_x^2 U_j^{n+1} + \delta_x^2 U_j^n + \delta_x^2 U_j^{n-1} \right\}$$

is stable for all values of μ.
(iii) Show that the scheme

$$U_j^{n+1} - U_j^{n-1} = \tfrac{1}{6}\mu \left\{ \delta_x^2 U_j^{n+1} + 4\delta_x^2 U_j^n + \delta_x^2 U_j^{n-1} \right\}$$

is unstable for all values of μ.

2.7 Find the leading terms in the truncation error of the explicit scheme

$$\frac{U_j^{n+1} - U_j^n}{\Delta t} = \frac{\{(U_{j+1}^n - U_j^n)p_{j+1/2} - (U_j^n - U_{j-1}^n)p_{j-1/2}\}}{(\Delta x)^2}$$

for the differential equation

$$\frac{\partial u}{\partial t} = \frac{\partial}{\partial x}\left(p(x)\frac{\partial u}{\partial x}\right)$$

on the region $0 < x < 1$, $t > 0$, with boundary conditions specifying the values of u at $x = 0$ and $x = 1$. Deduce a bound on the global error of the result in terms of bounds on the derivatives of u and p, under the condition $0 < p(x)\Delta t \leq \tfrac{1}{2}(\Delta x)^2$.

2.8 Apply the θ-method to the problem of the previous exercise, showing that the conditions required for the stable use of the Thomas algorithm will hold if $p(x) > 0$. Show also that a maximum principle will apply provided that $2\Delta t(1-\theta)p(x) \leq (\Delta x)^2$ for all x.

2.9   Consider application of the θ-method to approximate the equation $u_t = u_{xx}$ with the choice

$$\theta = \tfrac{1}{2} + \frac{(\Delta x)^2}{12\Delta t}.$$

Show that the resulting scheme is unconditionally stable, has a truncation error which is $O((\Delta t)^2 + (\Delta x)^2)$ and provides rather more damping for all Fourier modes that oscillate from time step to time step than does the Crank–Nicolson scheme. However, show that the mesh ratio $\Delta t/(\Delta x)^2$ must lie in the interval $[\tfrac{1}{6}, \tfrac{7}{6}]$ for the maximum principle to apply.

2.10   To solve the equation $u_t = u_{xx}$ suppose that we use a non-uniform mesh in the x-direction, the mesh points being given by

$$x_j = \frac{j^2}{J^2}, \quad j = 0, 1, 2, \dots, J.$$

By the change of variable $x = s^2$, write the equation

$$u_t = \frac{1}{2s}\frac{\partial}{\partial s}\left(\frac{1}{2s}\frac{\partial u}{\partial s}\right);$$

use a uniform mesh with $\Delta s = 1/J$, and apply the difference scheme of Exercise 7, with the additional factor $1/2s_j$ on the right-hand side. Show that the leading terms of the truncation error are

$$T_j^n = \tfrac{1}{2}\Delta t\, u_{tt} - \tfrac{1}{24}(\Delta s)^2\frac{1}{2s}\left[\left(\frac{1}{2s}u_{sss}\right)_s + \left(\frac{1}{2s}u_s\right)_{sss}\right]$$

and that this may be transformed into

$$T_j^n = \tfrac{1}{2}\Delta t\, u_{tt} - (\Delta s)^2(\tfrac{2}{3}u_{xxx} + \tfrac{1}{3}x u_{xxxx}).$$

Compare this with the leading terms of the truncation error obtained in Exercise 3.

2.11   Suppose that the Crank–Nicolson scheme is used for the solution of the equation $u_t = u_{xx}$, with boundary conditions $U_0^n = U_J^n = 0$ for all $n \geq 0$, and the initial condition $U_k^0 = 1$ for a fixed k with $0 < k < J$, $U_j^0 = 0, j \neq k$. Write $w_j = U_j^1$, and verify that w_j satisfies the recurrence relation

$$-\tfrac{1}{2}\mu w_{j-1} + (1+\mu)w_j - \tfrac{1}{2}\mu w_{j+1} = q_j, \quad w_0 = w_J = 0,$$

where $q_k = 1 - \mu$, $q_{k+1} = q_{k-1} = \tfrac{1}{2}\mu$, and $q_j = 0$ otherwise.

Suppose that the mesh is sufficiently fine, and that the point x_k is sufficiently far from the boundary that both k and $J - k$ are large. Explain why a good approximation to w_j may be written

$$w_j = Ap^{|j-k|}, \quad j \neq k,$$
$$w_k = A + B,$$

where $p = (1 + \mu - \sqrt{(1 + 2\mu)})/\mu$. Write down and solve two equations for the constants A and B, and show that $w_k = 2/\sqrt{(1 + 2\mu)} - 1$. Deduce that (i) $w_k < 1$ for all $\mu > 0$; (ii) $w_k > 0$ if and only if $\mu < \frac{3}{2}$; and (iii) $w_k \geq w_{k+1}$ if and only if $\mu \leq (7 - \sqrt{17})/4$.

2.12   Show that the (Hermitian) difference scheme

$$(1 + \tfrac{1}{12}\delta_x^2)(U^{n+1} - U^n) = \tfrac{1}{2}\mu\delta_x^2(U^{n+1} + U^n)$$
$$+ \tfrac{1}{2}\Delta t[f^{n+1} + (1 + \tfrac{1}{6}\delta_x^2)f^n]$$

for approximating $u_t = u_{xx} + f$, for a given function f and with fixed $\mu = \Delta t/(\Delta x)^2$, has a truncation error which is $O((\Delta t)^2)$.

3

Parabolic equations in two and three dimensions

3.1 The explicit method in a rectilinear box

The natural generalisation of the one-dimensional model problem in two dimensions is the equation

$$u_t = b\nabla^2 u \quad (b > 0)$$
$$= b\left[u_{xx} + u_{yy}\right], \tag{3.1}$$

where b is a positive constant. We shall consider the rectangular domain in the (x, y)-plane

$$0 < x < X, \quad 0 < y < Y,$$

and assume Dirichlet boundary conditions, so that $u(x, y, t)$ is given at all points on the rectangular boundary, for all positive values of t. In addition, of course, an initial condition is given, so that $u(x, y, 0)$ is given on the rectangular region. The region is covered with a uniform rectangular grid of points, with a spacing Δx in the x-direction and Δy in the y-direction, where

$$\Delta x = \frac{X}{J_x}, \quad \Delta y = \frac{Y}{J_y}, \quad J_x, J_y \in \mathbb{Z}.$$

The approximate solution is then denoted by

$$U_{r,s}^n \approx u(x_r, y_s, t_n), \quad r = 0, 1, \ldots, J_x, \quad s = 0, 1, \ldots, J_y.$$

The simplest explicit difference scheme is the natural extension of the explicit scheme in one dimension, and is given by

$$\frac{U^{n+1} - U^n}{\Delta t} = b\left[\frac{\delta_x^2 U^n}{(\Delta x)^2} + \frac{\delta_y^2 U^n}{(\Delta y)^2}\right]. \tag{3.2}$$

Here we have omitted the subscripts (r,s) throughout, and used the notation of (2.28) for the second order central differences in the x- and y-directions. This is an explicit scheme, since there is only one unknown value $U_{r,s}^{n+1}$ on the new time level. This unknown value is calculated from five neighbouring values on the previous time level,

$$U_{r,s}^n, \ U_{r+1,s}^n, \ U_{r-1,s}^n, \ U_{r,s+1}^n \text{ and } U_{r,s-1}^n. \tag{3.3}$$

Most of the analysis of this scheme in one dimension is easily extended to the two-dimensional case; the details are left as an exercise. The truncation error is

$$T(x,t) = \tfrac{1}{2}\Delta t\, u_{tt} - \tfrac{1}{12}b\left[(\Delta x)^2 u_{xxxx} + (\Delta y)^2 u_{yyyy}\right] + \cdots, \tag{3.4}$$

from which a bound on the truncation error can be obtained in terms of bounds on the derivatives of u, written in the same notation as before as M_{tt}, M_{xxxx} and M_{yyyy}. The proof of convergence follows in a similar way, leading to

$$E^n \le \left[\tfrac{1}{2}\Delta t\, M_{tt} + \tfrac{1}{12}b\left((\Delta x)^2 M_{xxxx} + (\Delta y)^2 M_{yyyy}\right)\right] t_F, \tag{3.5}$$

provided that the mesh sizes satisfy the condition

$$\mu_x + \mu_y \le \tfrac{1}{2}, \tag{3.6}$$

where

$$\mu_x = \frac{b\Delta t}{(\Delta x)^2}, \quad \mu_y = \frac{b\Delta t}{(\Delta y)^2}. \tag{3.7}$$

The stability of the scheme can also be analysed by Fourier series, assuming that b is constant, and ignoring the effect of boundary conditions, for which a justification will be given in Chapter 5. We construct solutions of the difference equation of the form

$$U^n \sim (\lambda)^n \exp\left(\mathrm{i}\left[k_x x + k_y y\right]\right) \tag{3.8}$$

to obtain the amplification factor λ as

$$\lambda \equiv \lambda(\mathbf{k}) = 1 - 4\left[\mu_x \sin^2 \tfrac{1}{2}k_x \Delta x + \mu_y \sin^2 \tfrac{1}{2}k_y \Delta y\right] \tag{3.9}$$

where $\mathbf{k} = (k_x, k_y)$. Just as in the one-dimensional problem it is clear that the stability condition is

$$\mu_x + \mu_y \le \tfrac{1}{2}. \tag{3.10}$$

The sufficiency of this condition in establishing $|\lambda(\mathbf{k})| \le 1$ follows just as in the one-dimensional case: its necessity follows from the fact that all the components of $\mathbf{k}$ can be chosen independently to give the worst mode, for which $k_x \Delta x = k_y \Delta y = \pi$.

Calculating the approximation from (3.2) is clearly just as easy as in the one-dimensional case. However the stability condition is more restrictive; and when b is variable we need to apply the condition (3.10) at each point so that any local peak in b will cut down the time step that can be used. Thus this simple explicit scheme is generally impractical and we must introduce some implicitness to avoid, or relax, the stability restriction. This is even more true in three dimensions, to which all of the above is readily extended.

3.2 An ADI method in two dimensions

The natural extension of our study of the one-dimensional problem would now be to suggest an extension of the θ-method . In particular, the Crank–Nicolson method becomes

$$\left(1 - \tfrac{1}{2}\mu_x \delta_x^2 - \tfrac{1}{2}\mu_y \delta_y^2\right) U^{n+1} = \left(1 + \tfrac{1}{2}\mu_x \delta_x^2 + \tfrac{1}{2}\mu_y \delta_y^2\right) U^n. \qquad (3.11)$$

In one dimension the great advantage of this type of method was the lifting of the stability restriction with little extra computational labour. In two or more dimensions this is no longer true; the method is still stable without restriction on the time step, but the extra labour involved is now very considerable. We have to solve a system of $(J_x - 1)(J_y - 1)$ linear equations for the unknown values $U_{r,s}^{n+1}$. The equations have a regular structure, each equation involving at most five unknowns; the matrix of the system consists very largely of zeros, but it does not have tridiagonal form; moreover there is no way of permuting the rows and columns so that the non-zero elements form a narrow band. The solution of such a system of equations is by no means out of the question, as we shall see when we come to elliptic equations in Chapter 6, but it requires so much extra sophistication that it suggests we should look for other numerical schemes for parabolic equations in two dimensions.

Since an implicit method in one dimension can be very efficient, it is natural to look for methods which are implicit in one dimension, but not both. Consider, for example, the scheme

$$\left(1 - \tfrac{1}{2}\mu_x \delta_x^2\right) U^{n+1} = \left(1 + \tfrac{1}{2}\mu_x \delta_x^2 + \mu_y \delta_y^2\right) U^n. \qquad (3.12)$$

If we examine the equations of the system corresponding to a particular row of mesh points or value of s, we see that they form a tridiagonal system of order $J_x - 1$, as they do not involve any unknowns with different values of s. The complete system thus involves a set of $J_y - 1$ tridiagonal systems, each of which can be solved very efficiently by the Thomas algorithm of Section 2.9. This scheme will require roughly three times as much computational labour as the explicit scheme. Unfortunately, although the stability of the scheme has been improved, there is still a restriction. We easily find, still assuming b is constant, that the amplification factor is

$$\lambda(\mathbf{k}) = \frac{1 - 2\mu_x \sin^2 \frac{1}{2} k_x \Delta x - 4\mu_y \sin^2 \frac{1}{2} k_y \Delta y}{1 + 2\mu_x \sin^2 \frac{1}{2} k_x \Delta x}$$

and the scheme will be unstable if $\mu_y > \frac{1}{2}$. As might be expected, there is no restriction on μ_x.

Successful methods can be obtained by combining two such schemes, each of which is implicit in one direction. The first such scheme was proposed by Peaceman and Rachford in 1955[1] and used by them in oil reservoir modelling. We begin by writing a modification of the Crank–Nicolson scheme in the form

$$\left(1 - \tfrac{1}{2}\mu_x\delta_x^2\right)\left(1 - \tfrac{1}{2}\mu_y\delta_y^2\right)U^{n+1} = \left(1 + \tfrac{1}{2}\mu_x\delta_x^2\right)\left(1 + \tfrac{1}{2}\mu_y\delta_y^2\right)U^n. \quad (3.13)$$

Noticing that we can expand the product of the difference operators as

$$\left(1 + \tfrac{1}{2}\mu_x\delta_x^2\right)\left(1 + \tfrac{1}{2}\mu_y\delta_y^2\right) = \left(1 + \tfrac{1}{2}\mu_x\delta_x^2 + \tfrac{1}{2}\mu_y\delta_y^2 + \tfrac{1}{4}\mu_x\mu_y\delta_x^2\delta_y^2\right),$$
$$(3.14)$$

we see that (3.13) is not exactly the same as the Crank–Nicolson scheme, but introduces extra terms which are of similar order to some in the truncation error – see (3.19) below. Introducing an intermediate level $U^{n+1/2}$, (3.13) can be written in the equivalent form

$$\left(1 - \tfrac{1}{2}\mu_x\delta_x^2\right)U^{n+1/2} = \left(1 + \tfrac{1}{2}\mu_y\delta_y^2\right)U^n, \quad (3.15a)$$
$$\left(1 - \tfrac{1}{2}\mu_y\delta_y^2\right)U^{n+1} = \left(1 + \tfrac{1}{2}\mu_x\delta_x^2\right)U^{n+1/2}, \quad (3.15b)$$

the equivalence being seen by operating on (3.15a) with $\left(1 + \tfrac{1}{2}\mu_x\delta_x^2\right)$ and on (3.15b) with $\left(1 - \tfrac{1}{2}\mu_x\delta_x^2\right)$.

In (3.15a) the terms on the right-hand side are known from the previous step; having computed $U^{n+1/2}$, the terms on the right-hand side of (3.15b) are then also known. Just as for the singly implicit scheme (3.12) the solution of each of the systems involves sets of tridiagonal equations.

[1] Peaceman, D.W. and Rachford, H.H. Jr (1955), The numerical solution of parabolic and elliptic differential equations, *J. Soc. Indust. Appl. Math.* **3**, 28.

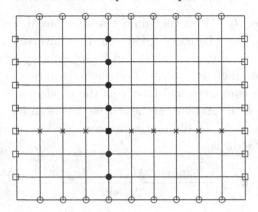

Fig. 3.1. Boundary points for an ADI method.

The total work involved in one time step amounts to solving $J_y - 1$ tridiagonal systems (for points such as those marked by crosses in Fig. 3.1), each of order $J_x - 1$, followed by solving $J_x - 1$ similar systems (for points such as those marked by dots in Fig. 3.1), each of order $(J_y - 1)$. This whole process can be carried out very much faster than the solution of the full system of order $(J_x - 1)(J_y - 1)$ in the Crank–Nicolson method; we need approximately 10(add)+8(mult)+6(div) operations per mesh point as compared with 4(add)+3(mult) for the explicit scheme (3.2): that is about three times as much computation. Boundary conditions for $U^{n+1/2}$ are needed at all the points marked with $\square$ in Fig. 3.1 and for U^n at the points marked with $\odot$.

When b is constant we can again analyse the stability of (3.13) by substituting the Fourier mode (3.8). From either form we obtain

$$\lambda(\mathbf{k}) = \frac{\left(1 - 2\mu_x \sin^2 \frac{1}{2} k_x \Delta x\right)\left(1 - 2\mu_y \sin^2 \frac{1}{2} k_y \Delta y\right)}{\left(1 + 2\mu_x \sin^2 \frac{1}{2} k_x \Delta x\right)\left(1 + 2\mu_y \sin^2 \frac{1}{2} k_y \Delta y\right)} \qquad (3.16)$$

from which the scheme's unconditional stability follows immediately.

We can also apply maximum principle arguments to (3.15). In the first half-step an individual equation takes the form

$$(1+\mu_x)U_{r,s}^{n+1/2} = (1-\mu_y)U_{r,s}^n + \tfrac{1}{2}\mu_y\left(U_{r,s-1}^n + U_{r,s+1}^n\right)$$
$$+ \tfrac{1}{2}\mu_x\left(U_{r+1,s}^{n+1/2} + U_{r-1,s}^{n+1/2}\right). \qquad (3.17)$$

Thus, provided that $\mu_y \leq 1$, the value of $U_{r,s}^{n+1/2}$ is expressed as a linear combination, with nonnegative coefficients summing to unity, of neighbouring values of U^n and $U^{n+1/2}$. The same is evidently also true of the equation in the second half-step. Thus the maximum principle shows that the numerical values are bounded by the maximum and minimum values on the boundaries, provided that

$$\max\{\mu_x, \mu_y\} \leq 1 \tag{3.18}$$

which is a natural generalisation of the condition for the one-dimensional Crank–Nicolson method.

The truncation error is most readily calculated from the unsplit form (3.13). Taking all terms to the left and dividing by Δt we fairly readily deduce that the leading terms are

$$T^{n+1/2} \approx \tfrac{1}{24}(\Delta t)^2 u_{ttt} - \tfrac{1}{12}(\Delta x)^2 u_{xxxx} - \tfrac{1}{12}(\Delta y)^2 u_{yyyy}$$
$$-\tfrac{1}{8}(\Delta t)^2 u_{xxtt} - \tfrac{1}{8}(\Delta t)^2 u_{yytt} + \tfrac{1}{4}(\Delta t)^2 u_{xxyyt} \tag{3.19}$$
$$= O\left((\Delta t)^2 + (\Delta x)^2 + (\Delta y)^2\right),$$

the first five terms being as for the Crank–Nicolson scheme in two dimensions (cf. (2.85) for the one-dimensional case) and the last coming from the product term $\delta_x^2 \delta_y^2 \left(U^{n+1} - U^n\right)$.

As an example we consider the heat equation

$$u_t = u_{xx} + u_{yy} \tag{3.20}$$

on the unit square $0 < x < 1$, $0 < y < 1$, with homogeneous Dirichlet boundary conditions $u = 0$ on the boundary of the unit square. The initial condition is $u(x, y, 0) = f(x, y)$, where $f(x, y) = 1$ within the region shaped like the letter **M**, as in Fig. 3.2, and $f(x, y) = 0$ in the rest of the square. In a narrow band surrounding the **M**, the function increases linearly from 0 to 1, so that $f(x, y)$ is continuous; its derivatives are not continuous, being zero everywhere outside this narrow band and being large inside the band.

Results of an explicit numerical calculation are shown in Fig. 3.3, Fig. 3.4 and Fig. 3.5; they illustrate the way in which this initial function diffuses throughout the square. It is possible in principle to determine the error in the numerical solution by comparison with a Fourier series expansion of the exact solution, as in Chapter 2. The coefficients in this Fourier series would be extremely complicated, and instead we have estimated the accuracy of the numerical solution by repeating the calculations with different mesh sizes. Using $\Delta x = \Delta y = 1/100$, $1/200$ and $1/400$ indicated good agreement, and gives us confidence that the solutions illustrated in the graphs are more accurate than the resolution available in the graphical representation.

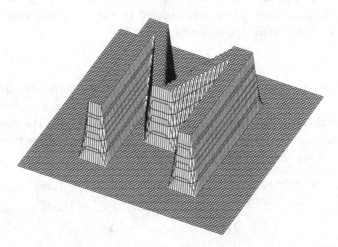

Fig. 3.2. Initial data for a heat flow calculation on the unit square.

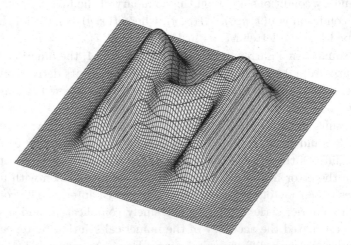

Fig. 3.3. The numerical solution at $t = 0.001$ for the data of Fig. 3.2.

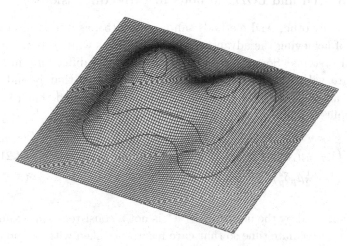

Fig. 3.4. As for Fig. 3.3 at $t = 0.004$.

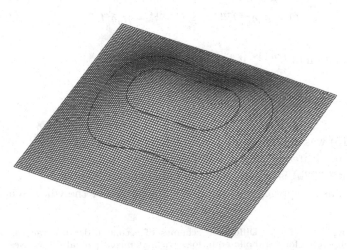

Fig. 3.5. As for Fig. 3.3 at $t = 0.01$.

3.3 ADI and LOD methods in three dimensions

There are many other ADI methods similar to (3.15), as well as alternative ways of achieving the advantages of implicitness while solving only tridiagonal systems. Sometimes variants of a scheme differ only in the intermediate values calculated, which affects the ways that boundary conditions are imposed. For example D'yakonov[1] proposed that (3.13) could be split as follows :

$$\left(1 - \tfrac{1}{2}\mu_x\delta_x^2\right) U^{n+*} = \left(1 + \tfrac{1}{2}\mu_x\delta_x^2\right)\left(1 + \tfrac{1}{2}\mu_y\delta_y^2\right) U^n, \quad (3.21\text{a})$$

$$\left(1 - \tfrac{1}{2}\mu_y\delta_y^2\right) U^{n+1} = U^{n+*}. \quad (3.21\text{b})$$

Then, as indicated by the notation, U^{n+*} is not a consistent approximation at an intermediate time so that care has to be taken with boundary conditions: but these are needed for U^{n+*} only on the left and right (marked with $\square$ in Fig. 3.1) and can be obtained from (3.21b) at these points. Some of these variants have advantages when one considers generalisations to three dimensions.

A commonly used ADI scheme is due to Douglas and Rachford[2] which in two dimensions takes the form

$$\left(1 - \mu_x\delta_x^2\right) U^{n+1*} = \left(1 + \mu_y\delta_y^2\right) U^n, \quad (3.22\text{a})$$

$$\left(1 - \mu_y\delta_y^2\right) U^{n+1} = U^{n+1*} - \mu_y\delta_y^2 U^n; \quad (3.22\text{b})$$

after elimination of U^{n+1*} it leads to

$$\left(1 - \mu_x\delta_x^2\right)\left(1 - \mu_y\delta_y^2\right) U^{n+1} = \left(1 + \mu_x\mu_y\delta_x^2\delta_y^2\right) U^n. \quad (3.23)$$

While (3.15) was motivated by the Crank–Nicolson scheme, (3.22, 3.23) is clearly motivated more by the fully implicit scheme. It is readily shown that (3.22b) is consistent with the differential equation if U^{n+1*} is regarded as an approximation to u^{n+1} and that for the whole scheme

[1] D'yakonov, E.G. (1964), Difference schemes of second order accuracy with a splitting operator for parabolic equations without mixed partial derivatives, *Zh. Vychisl. Mat. i Mat. Fiz.*, **4**, 935–41.

[2] Douglas, J. Jr and Rachford, H.H. Jr (1956), On the numerical solution of the heat conduction problems in two and three variables, *Trans. Amer. Math. Soc.* **82**, 421–39.

the truncation error is $O(\Delta t + (\Delta x)^2)$. It is also easily generalised to three dimensions as

$$\left(1 - \mu_x \delta_x^2\right) U^{n+1*} = \left(1 + \mu_y \delta_y^2 + \mu_z \delta_z^2\right) U^n, \tag{3.24a}$$

$$\left(1 - \mu_y \delta_y^2\right) U^{n+1**} = U^{n+1*} - \mu_y \delta_y^2 U^n, \tag{3.24b}$$

$$\left(1 - \mu_z \delta_z^2\right) U^{n+1} = U^{n+1**} - \mu_z \delta_z^2 U^n. \tag{3.24c}$$

The same remarks as above apply to this.

A Fourier analysis applied to (3.22), (3.23) or (3.24) easily shows that they are unconditionally stable, as in the one-dimensional case. However, maximum principle arguments are much more difficult to apply in either the form (3.22) or (3.23).

An alternative, much favoured by Russian authors, is to use the locally one-dimensional (LOD) schemes in which only one variable is dealt with at a time. Thus based on the Crank–Nicolson scheme we have, using two intermediate levels,

$$\left(1 - \tfrac{1}{2}\mu_x \delta_x^2\right) U^{n+*} = \left(1 + \tfrac{1}{2}\mu_x \delta_x^2\right) U^n, \tag{3.25a}$$

$$\left(1 - \tfrac{1}{2}\mu_y \delta_y^2\right) U^{n+**} = \left(1 + \tfrac{1}{2}\mu_y \delta_y^2\right) U^{n+*}, \tag{3.25b}$$

$$\left(1 - \tfrac{1}{2}\mu_z \delta_z^2\right) U^{n+1} = \left(1 + \tfrac{1}{2}\mu_z \delta_z^2\right) U^{n+**}. \tag{3.25c}$$

Since the intermediate values are not at all consistent with the full differential equation, special care needs to be taken with their boundary conditions. For example, suppose we denote by V^n, V^{n+*} the values obtained by the equivalent of (3.25) in two dimensions and compare with the Peaceman–Rachford scheme (3.15). Eliminating V^{n+*} we have

$$\left(1 - \tfrac{1}{2}\mu_y \delta_y^2\right) V^{n+1} = \left(1 + \tfrac{1}{2}\mu_y \delta_y^2\right) \left(1 - \tfrac{1}{2}\mu_x \delta_x^2\right)^{-1} \left(1 + \tfrac{1}{2}\mu_x \delta_x^2\right) V^n \tag{3.26}$$

from which by comparison with (3.13) we deduce that the LOD scheme is equivalent to the Peaceman–Rachford scheme if we write

$$V^n = \left(1 - \tfrac{1}{2}\mu_x \delta_x^2\right) U^n, \quad V^{n+*} = \left(1 + \tfrac{1}{2}\mu_x \delta_x^2\right) U^n. \tag{3.27}$$

This relationship can be exploited in various ways to obtain reasonably good boundary conditions.

3.4 Curved boundaries

Another important new problem in two or more dimensions is that the region will usually be more complicated. In one dimension it is necessary to consider only a finite interval, which can be standardised to be

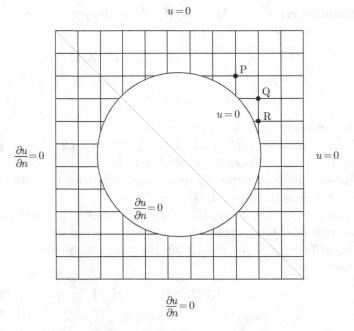

Fig. 3.6. Model problem with curved boundary.

the interval $[0, 1]$. So far in two dimensions we have considered only the natural extension of this, which is the unit square; but in general the region will have curved boundaries, and may well not be simply connected. An obvious example of physical interest could be a square plate with a round hole in it. Some of the difficulties which now arise will be mainly computational, concerning the way in which the calculation is organised; for example, if we cover a general region with a square or rectangular grid the numbers of grid points along the grid lines will no longer be all the same. Here we shall not consider such issues, but we must discuss the way in which the boundary conditions are incorporated into the finite difference equations.

We take as an example the problem illustrated in Fig. 3.6. The region Ω lies between the unit square and the circle with centre at $(\frac{1}{2}, \frac{1}{2})$ and radius 0.33 which form the boundary $\partial\Omega$; we wish to solve the heat equation

$$u_t = u_{xx} + u_{yy} \tag{3.28}$$

in this region. On the part of the region above the line $x + y = 1$ the boundary condition specifies that $u = 0$ on the two straight sides of the region, and also on this half of the interior circle. On the other parts of the boundary, where $x + y < 1$, the boundary condition specifies that the normal derivative is zero. Thus

$$u(x, y, t) = 0 \text{ on } \{x + y \geq 1\} \cap \partial\Omega, \tag{3.29}$$

$$\frac{\partial u}{\partial n} = 0 \text{ on } \{x + y < 1\} \cap \partial\Omega. \tag{3.30}$$

The initial condition requires that $u(x, y, 0) = 0$ except in two small symmetrically placed regions near the North-East and South-West corners. Physically this problem models the flow of heat in a square plate with a round hole in it. Half of the boundary of the plate is maintained at zero temperature, and the other half is insulated; initially the plate is at zero temperature, except for two hot spots.

As shown in Fig. 3.6 the square is covered with a uniform square grid of size $\Delta x = \Delta y = 1/J$. At many of the points of this grid the standard difference approximations are used, but when the points are near the boundaries we need to apply some special formulae. Consider the point P; its two neighbours on the East and West are ordinary grid points, and there is no difficulty about $\delta_x^2 U$, but its neighbour on the South would be inside the circle. The situation is shown enlarged in Fig. 3.7.

Here we need an approximation to u_{yy} which uses values at P and N, and at the boundary point B. Such an approximation is quite easily constructed; it generalises the standard form involving three neighbouring points to the situation in which these points are not equally spaced.

There are various ways in which the result can be derived. One of the most straightforward is to write

$$\frac{u_N - u_P}{y_N - y_P} \approx \frac{\partial u}{\partial y}(P_+), \quad \frac{u_P - u_B}{y_P - y_B} \approx \frac{\partial u}{\partial y}(P_-)$$

as approximations to the first partial derivative at the two points P_+ and P_-, one half-way between P and N, and one half-way between P and B. Since the distance between these two half-way points is $(y_N - y_B)/2$, we obtain the required approximation

$$u_{yy} \approx \frac{2}{y_N - y_B} \left(\frac{u_N - u_P}{y_N - y_P} - \frac{u_P - u_B}{y_P - y_B} \right). \tag{3.31a}$$

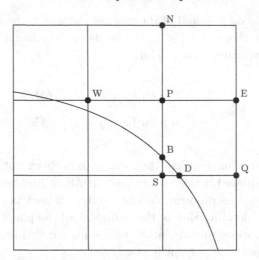

Fig. 3.7. Dirichlet condition on curved boundary.

In this example $y_P - y_S = \Delta y$, and we can write $y_P - y_B = \alpha \Delta y$, where $0 < \alpha < 1$. The approximation then becomes

$$u_{yy} \approx \frac{2}{(\alpha+1)(\Delta y)^2} u_N - \frac{2}{\alpha(\Delta y)^2} u_P + \frac{2}{\alpha(\alpha+1)(\Delta y)^2} u_B. \quad (3.31b)$$

A similar approximation is used at a point like Q, where one of the neighbours is replaced by the point D on the boundary, and we need to approximate the derivative u_{xx}. There will also be points like R where two of the neighbours are outside the region, and both x- and y- derivatives need special treatment.

It is worth noting here that these formulae are equivalent to extrapolating from interior and boundary points to a fictitious exterior point, and then applying the standard difference scheme. For example, the quadratic extrapolation

$$u_S = \frac{\alpha(1-\alpha)u_N + 2u_B - 2(1-\alpha^2)u_P}{\alpha(\alpha+1)} \quad (3.32)$$

would lead to (3.31b).

If we are using the explicit method, in general we may obtain a difference scheme of the form

$$\frac{U_{r,s}^{n+1} - U_{r,s}^n}{\Delta t} = \frac{2}{(1+\alpha)(\Delta x)^2} U_{r+1,s}^n + \frac{2}{\alpha(1+\alpha)(\Delta x)^2} u_B$$

$$+ \frac{2}{(1+\beta)(\Delta y)^2} U_{r,s+1}^n + \frac{2}{\beta(1+\beta)(\Delta y)^2} u_D$$

$$- \left(\frac{2}{\alpha(\Delta x)^2} + \frac{2}{\beta(\Delta y)^2} \right) U_{r,s}^n, \tag{3.33}$$

where u_B and u_D are given values of the solution on the boundary. The expression for $U_{r,s}^{n+1}$, in terms of the values on the previous time level, thus retains the important property that the coefficients of the neighbouring values are all nonnegative, provided that the time step Δt is not too large. The stability restriction for the explicit method is similar to the standard case, and the error analysis is largely unaltered. However, the restriction on the time step is likely to be much more serious than for the rectangular boundary, since the condition is now

$$2 \left(\frac{\mu_x}{\alpha} + \frac{\mu_y}{\beta} \right) \leq 1; \tag{3.34}$$

α and β are both less than 1 and may be quite small. For example, for the present problem when we use $\Delta x = \Delta y = 1/50$ we find that the maximum permitted time step is 0.000008; while without the hole the maximum time step would be 0.00001, which is not very different. However, if we double the number of mesh points in both the x- and y-directions the maximum time step without the hole is reduced by a factor of 4; while with the hole this puts one of the mesh points rather close to the boundary, and the time step is reduced by a factor of 800, to 10^{-7}. Such problems drive home the requirement to use implicit methods, or more appropriate meshes – see below.

Normal derivative boundary conditions can be handled in a similar way but are considerably more difficult. For example, in Fig. 3.8 suppose that we are given the outward normal derivative at the point B. The normal at B meets the horizontal mesh line WPE at Z and suppose the lengths are

$$ZP = p\Delta x, \quad PB = \alpha\Delta y, \quad BZ = q\Delta y,$$

where $0 \leq p \leq 1$, $0 < \alpha \leq 1$ and $0 < q \leq \sqrt{[1 + (\Delta x)^2/(\Delta y)^2]}$. The normal derivative can be approximated by

$$\frac{u_B - u_Z}{q\Delta y} \approx \frac{\partial u}{\partial n} = g(B), \tag{3.35}$$

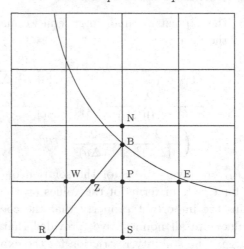

Fig. 3.8. Neumann condition on curved boundary.

and the value of u_Z can be obtained by linear interpolation between u_W and u_P,

$$u_Z \approx p u_W + (1-p) u_P. \tag{3.36}$$

As before we approximate u_{xx} in terms of u_W, u_P, u_E, and u_{yy} in terms of u_S, u_P, u_B; we then eliminate u_B and u_Z to give the scheme

$$
\begin{aligned}
\frac{U_P^{n+1} - U_P^n}{\Delta t} &= \frac{U_E - 2U_P + U_W}{(\Delta x)^2} \\
&\quad + \frac{1}{(\Delta y)^2}\left\{ \frac{2}{\alpha(\alpha+1)}U_B - \frac{2}{\alpha}U_P + \frac{2}{\alpha+1}U_S \right\} \\
&= \frac{U_E - 2U_P + U_W}{(\Delta x)^2} + \frac{1}{(\Delta y)^2}\left\{ -\frac{2}{\alpha}U_P + \frac{2}{\alpha+1}U_S \right\} \\
&\quad + \frac{2}{\alpha(\alpha+1)(\Delta y)^2}\left\{ pU_W + (1-p)U_P + qg_B\Delta y \right\} \\
&= \frac{1}{(\Delta x)^2}U_E + \left\{ \frac{1}{(\Delta x)^2} + \frac{2p}{\alpha(\alpha+1)(\Delta y)^2} \right\}U_W \\
&\quad + \frac{2}{(\alpha+1)(\Delta y)^2}U_S - \left\{ \frac{2}{(\Delta x)^2} + \frac{2(\alpha+p)}{\alpha(\alpha+1)(\Delta y)^2} \right\}U_P \\
&\quad + \frac{2q}{\alpha(\alpha+1)\Delta y}g_B. \tag{3.37}
\end{aligned}
$$

Notice how once again this expression involves positive coefficients for the values of U at the neighbouring points. The same expressions can be incorporated into a Crank–Nicolson or ADI scheme, for which a maximum principle will hold with appropriate limits on Δt. We see again, by equating the right-hand side of (3.37) to the standard scheme using U_N and deducing an extrapolation formula for U_N, that the scheme is equivalent to extrapolation for U_N first and then applying the standard formula.

The various forms of ADI method are very little changed in these situations; the main difference in practice is the extra calculation involved in forming the elements of the tridiagonal matrices near the boundary, and the fact that these matrices will now have different sizes.

These methods for approximating the curved boundary lead to truncation errors of lower order than those at ordinary internal points, especially where normal derivatives are involved. Just as we found in the one-dimensional case in Section 2.13, the truncation error may not tend to zero at these points when the mesh is refined. It is difficult to produce higher order approximations with the desirable properties. For example, we have used the simplest approximation to the normal derivative, incorporating just the two points U_B and U_Z. Now suppose we extend the normal at B to the point R in Fig. 3.8, where $ZR = (q/\alpha)\Delta y$. Then a higher order approximation to the derivative is

$$g_B = \frac{\partial u}{\partial n} \approx \frac{(1 + 2\alpha)u_B + \alpha^2 u_R - (1 + \alpha)^2 u_Z}{(1 + \alpha)q\Delta y}. \qquad (3.38)$$

It will now be more awkward to interpolate for the value u_R, and moreover the coefficients of u_Z and u_R have opposite signs. The resulting scheme will not satisfy a maximum principle. It seems likely that such a higher order scheme will be unsatisfactory in practice unless further restrictions are imposed.

The results of a typical calculation using the Peaceman–Rachford ADI method are shown in Figs. 3.9–3.12. These used a grid of size 50 in the x- and y-directions. The diagrams are not to the same scale, and correspond to an approximately North-East viewing point so that the smaller peak is in the foreground; the two initial peaks in u diffuse over the region, and become much smaller, so each figure is scaled to show the same maximum height. The actual maximum value of u is also given. The results show clearly how one of the peaks diffuses through the boundary, while the other spreads round the hole, as it cannot

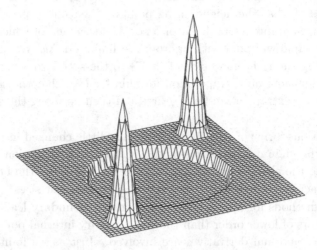

Fig. 3.9. Initial data for a heat flow calculation on the
domain of Fig. 3.6; maximum $u = 1$.

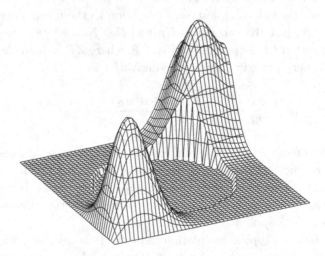

Fig. 3.10. Numerical solution at $t = 0.005$ for the
data of Fig. 3.9; maximum $U = 0.096$.

diffuse through the insulated part of the boundary. Eventually it diffuses
round the hole and out of the other half of the boundary. In Fig. 3.11
this is beginning to happen, and the other peak has almost completely
disappeared.

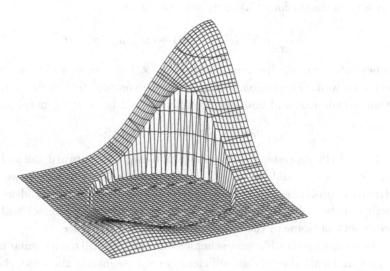

Fig. 3.11. As for Fig. 3.10 at $t = 0.02$; maximum $U = 0.060$.

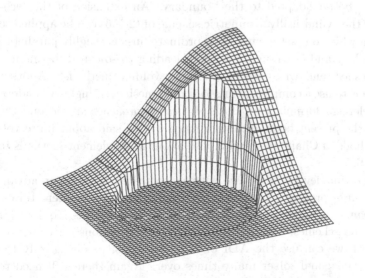

Fig. 3.12. As for Fig. 3.10 at $t = 0.05$; maximum $U = 0.038$.

3.5 Application to general parabolic problems

In several dimensions (3.1) may take the form

$$e\frac{\partial u}{\partial t} = \nabla \cdot (b\nabla u - \mathbf{a}u) + cu + d, \quad \mathbf{x} \in \Omega, \tag{3.39}$$

where all the coefficients may depend on $\mathbf{x}$, t and in nonlinear problems on u as well. The region Ω may be quite complicated in shape rather than rectilinear, and boundary conditions will be of the general form

$$\alpha_0 u + \alpha_1 \partial u/\partial n = \alpha_2 \text{ on } \partial\Omega \tag{3.40}$$

where $\partial/\partial n$ represents differentiation along the outward normal and $\alpha_0 \geq 0$, $\alpha_1 \geq 0$ and $\alpha_0 + \alpha_1 > 0$. As indicated in Chapter 1, we have tried to emphasise in the above account those methods and ideas that will still be valid for such more general problems. Here we will just comment on some of the extra difficulties that will arise.

In setting up the difference schemes we can often still use a regular mesh, and with local values of the coefficients we use essentially the same schemes as given above. The main problems occur at curved boundaries where the boundary conditions have to be applied using an irregularly arranged set of points; and both the variation in mesh size that results and the variation of coefficients make it more necessary to use implicit methods.

Alternative approaches may be based on the use of irregular meshes which are better adapted to the boundary. An extension of the technique for the cylindrically symmetric scheme of (2.156) can be applied to a mesh in which one set of curved co-ordinate lines is roughly parallel to the boundary and one roughly normal, leading to so-called 'boundary-fitted meshes' and what are called finite volume methods. Another approach is to use a completely unstructured mesh of triangles and adopt a finite element formulation. Both of these approaches are beyond the scope of the present book, although more will be said about finite volume methods in Chapters 4 and 5, and about finite element methods in Chapter 6.

The two-time-level schemes we have concentrated upon are an advantage for large problems because they minimise the computer storage requirements: not much more than U at one time level is required for most of the schemes. To solve the difference equations on a rectangular mesh we can use the ADI or LOD methods, requiring only the use of a tridiagonal solver many times over. Again there will need to be adjustments at the boundary in order to obtain accurate boundary conditions; this will be particularly true at re-entrant corners. These remarks assume that the equations are linear. If they are nonlinear in not too severe a way, we can linearise about the earlier time level so that implicit methods may be used, and still hope to obtain reasonable accuracy. So far as the analysis of the scheme's behaviour is concerned we cannot of course carry out a complete Fourier analysis unless the scheme is linear, has constant coefficients and is in a rectangular region with a uniform mesh and periodic boundary conditions. However, one can show that the stability conditions obtained by a localised Fourier analysis are still necessary: this is essentially because the most unstable modes in all the cases we have encountered are of high frequency and therefore grow locally. Thus for linear problems we substitute Fourier modes into the difference equations with the coefficients held constant at local values in order to seek out unstable solutions: for nonlinear problems we do the same to the linearised equations. These topics will be discussed further in Chapter 5.

To establish convergence for linear problems the sharpest tool has to be the maximum principle. Provided a correct choice of difference scheme is used, particularly for problems involving derivative boundary conditions, a maximum principle will still hold, but we must expect gaps between the conditions where we can prove convergence and when we can demonstrate instability. This will be particularly true with general derivative boundary conditions on a curved boundary. In particular, the analysis of ADI methods is more complicated for these problems. We have so far relied on the equality of the two factorisations

$$\left(1 - \tfrac{1}{2}\mu_x\delta_x^2\right)\left(1 - \tfrac{1}{2}\mu_y\delta_y^2\right)U = \left(1 - \tfrac{1}{2}\mu_y\delta_y^2\right)\left(1 - \tfrac{1}{2}\mu_x\delta_x^2\right)U,$$

because the two difference operators commute. At the irregular grid points near the boundaries this type of factorisation is only approximate, and the missing terms are not the same when the order of factorisation is reversed. The performance of these methods may then deteriorate, reflecting the fact that the Fourier analysis that established their key properties is no longer valid.

As we shall discuss in Chapter 7, during the last 20 to 30 years there have been enormous developments in methods and software to solve large systems of algebraic equations – particularly in *multigrid methods* following the work of Brandt.[1] These allow one to adopt the viewpoint that, for an implicit approximation to a parabolic problem, one has to solve a succession of elliptic problems, with each set of data provided

[1] Brandt, A. (1977) Multilevel adaptive solutions to boundary-value problems. *Math. Comp.* **31**, 333–90.

from the previous time step. So the problems of this chapter may be treated by the methods described in Chapters 6 and 7. Then ADI and LOD methods may play no role, or a very subsidiary role, as preconditioners for the algebraic solvers. If finite element methods are used, this will also mean that variation principles will provide the framework for analysing the methods.

So far we have not commented on the effect of the form of the equation (3.39). Inclusion of the term in **a**, which corresponds to convective effects in contrast to the diffusive effects arising from the leading term, will often make the equations less stable and we shall study this effect in the next two chapters and in Section 6.8. We have assumed that the diffusion coefficient b is a scalar and this is partly why so many of the methods and so much of the analysis have generalised easily. We could introduce a diagonal tensor, so that the diffusion coefficient is different in each of the co-ordinate directions, and this would also make very little change; but such anisotropy in practical applications is unlikely to line up neatly with our co-ordinate axes as it will represent real physical effects in, for example, stratified rock. Then the diffusion tensor has off-diagonal terms, and the equation will look like

$$eu_t = b_{11}u_{xx} + 2b_{12}u_{xy} + b_{22}u_{yy} - a_1u_x - a_2u_y + cu + d \qquad (3.41)$$

with a mixed second order derivative; and our difference schemes in two dimensions will involve nine rather than five points and twenty-seven in three dimensions.

These cause very much greater difficulties, and a proper discussion is beyond the scope of this book; but some remarks regarding nine-point schemes should be made here. Suppose the mixed derivative u_{xy} is approximated on a square mesh by the divided difference $\Delta_{0x}\Delta_{0y}U/(\Delta x)^2$. This will always give a negative coefficient at one of the points NE, NW, SE or SW in the compass-point notation of Figs. 3.7 and 3.8 and so invalidate the maximum principle. However, diagonal second differences could be used to approximate the u_{xx} and u_{yy} terms and these will make positive contributions to these coefficients. For example, suppose that in (3.41) we have $b_{11} > b_{22}$: then we can approximate $b_{22}(u_{xx} + u_{yy})$ by second differences in the two diagonal directions, providing positive contributions equal to $b_{22}/2(\Delta x)^2$. This will then ensure that the maximum principle is satisfied if $b_{22} \geq |b_{12}|$. Note, though, that this does not provide a remedy in all cases since the operator in (3.41) is elliptic provided only that $b_{11}b_{22} > b_{12}^2$. As this observation illustrates, the most difficult practical cases occur when diffusion is much greater in one direction than in others.

Bibliographic notes and recommended reading

The further reading recommended for Chapter 2 is equally relevant for this chapter, as well as much of that recommended in Chapter 6 for elliptic problems and in Chapter 7 for solving the linear systems of equations arising from their discretisation.

ADI and related methods are discussed fully in the book by Mitchell and Griffiths (1980); and many of the contributions to the subject from Russian authors are well covered in the book by Yanenko (1971).

Exercises

3.1   Show that the truncation error of the explicit scheme

$$\frac{U^{n+1} - U^n}{\Delta t} = b\left[\frac{\delta_x^2 U^n}{(\Delta x)^2} + \frac{\delta_y^2 U^n}{(\Delta y)^2}\right]$$

for the solution of the equation $u_t = b(u_{xx} + u_{yy})$, where $b = b(x, y) > 0$, has the leading terms

$$T^n = \tfrac{1}{2}\Delta t u_{tt} - \tfrac{1}{12}b[(\Delta x)^2 u_{xxxx} + (\Delta y)^2 u_{yyyy}].$$

If the boundary conditions prescribe the values of $u(x, y, t)$ at all points on the boundary of the rectangle $[0, X] \times [0, Y]$ for all $t > 0$, and the resulting solution is sufficiently smooth, show that the error in the numerical solution satisfies a bound, over $0 \le t \le t_F$, of the form

$$|U_{rs}^n - u(x_r, y_s, t_n)| \le t_F\{\tfrac{1}{2}\Delta t M_1 + \tfrac{1}{12}[(\Delta x)^2 M_2 + (\Delta y)^2 M_3]\}$$

provided that $b\Delta t[(\Delta x)^{-2} + (\Delta y)^{-2}] \le \tfrac{1}{2}$; identify the constants M_1, M_2 and M_3.

3.2   Show that the leading terms in the truncation error of the Peaceman–Rachford ADI method for $u_t = \nabla^2 u$ are

$$T^{n+1/2} = (\Delta t)^2[\tfrac{1}{24}u_{ttt} - \tfrac{1}{8}(u_{xxtt} + u_{yytt}) + \tfrac{1}{4}u_{xxyyt}]$$

$$-\tfrac{1}{12}[(\Delta x)^2 u_{xxxx} + (\Delta y)^2 u_{yyyy}].$$

How is this changed when a variable diffusivity b is introduced?

3.3   Show that the Douglas–Rachford scheme

$$(1 - \mu_x \delta_x^2)U^{n+1*} = (1 + \mu_y \delta_y^2 + \mu_z \delta_z^2)U^n,$$
$$(1 - \mu_y \delta_y^2)U^{n+1**} = U^{n+1*} - \mu_y \delta_y^2 U^n,$$
$$(1 - \mu_z \delta_z^2)U^{n+1} = U^{n+1**} - \mu_z \delta_z^2 U^n$$

for the heat equation $u_t = u_{xx} + u_{yy} + u_{zz}$ in three dimensions
has unrestricted stability when applied on a rectilinear box.

3.4   The explicit method on a uniform mesh is used to solve the
heat equation $u_t = u_{xx} + u_{yy}$ in a two-dimensional region with
a curved boundary; Dirichlet boundary conditions are given at
all points of the boundary. At points near the boundary the
scheme is modified to a form such as

$$\frac{U_{rs}^{n+1} - U_{rs}^n}{\Delta t} = \frac{2}{(1+\alpha)(\Delta x)^2}U_{r+1,s}^n + \frac{2}{\alpha(1+\alpha)(\Delta x)^2}U_B^n$$
$$+ \frac{2}{(1+\beta)(\Delta y)^2}U_{r,s+1}^n + \frac{2}{\beta(1+\beta)(\Delta y)^2}U_D^n$$
$$- \left(\frac{2}{\alpha(\Delta x)^2} + \frac{2}{\beta(\Delta y)^2}\right)U_{rs}^n,$$

where B and D are points on the boundary which are neighbours
of (x_r, y_s) and the distances from these points to (x_r, y_s) are
$\alpha\Delta x$ and $\beta\Delta y$. Find the leading terms in the truncation error
at this point, and show that the truncation error everywhere
has a bound of the form

$$|T_{rs}^n| \leq T := \tfrac{1}{2}\Delta t M_{tt} + \tfrac{1}{3}(\Delta x M_{xxx} + \Delta y M_{yyy})$$
$$+ \tfrac{1}{12}[(\Delta x)^2 M_{xxxx} + (\Delta y)^2 M_{yyyy}].$$

State the restriction on the size of Δt required for a maximum
principle to apply, and show that, if this condition is satisfied,
the error in the solution over the range $0 \leq t \leq t_F$ is bounded
by $t_F T$.

3.5   The explicit method on a uniform mesh is used to solve the
heat equation $u_t = u_{xx} + u_{yy}$ in the square region $0 \leq x \leq 1$,
$0 \leq y \leq 1$. The boundary conditions specify that $u_x = 0$
on the side $x = 0$ of the square, and Dirichlet conditions are
given on the rest of the boundary. At mesh points on $x = 0$
an additional line of mesh points with $x = -\Delta x$ is included,
and the extra values $U_{-1,s}^n$ are then eliminated by use of the

boundary condition. The scheme becomes, at points on the boundary $x = 0$,

$$\frac{U_{0,s}^{n+1} - U_{0,s}^n}{\Delta t} = \frac{2}{(\Delta x)^2}(U_{1,s}^n - U_{0,s}^n) + \frac{1}{(\Delta y)^2}\delta_y^2 U_{0,s}^n.$$

Show that the leading terms of the truncation error at this mesh point are

$$T_{0,s}^{n*} = \tfrac{1}{2}\Delta t u_{tt} - \tfrac{1}{12}[(\Delta x)^2 u_{xxxx} + (\Delta y)^2 u_{yyyy}] - \tfrac{1}{3}\Delta x u_{xxx}$$

and deduce that if the usual stability condition is satisfied the error in the solution satisfies $e_{r,s}^n \equiv U_{r,s}^n - u(x_r, y_s, t_n) = O(\Delta t) + O(\Delta x) + O((\Delta y)^2)$.

3.6   The diffusion equation $u_t = b\nabla^2 u$ on a rectangular region is to be approximated by a fractional-step method based on the Crank–Nicolson scheme over a uniform square mesh. Show that if b is constant the Peaceman–Rachford ADI method is exactly equivalent to the LOD method

$$(1 - \tfrac{1}{2}\mu\delta_y^2)V^{n+\frac{1}{2}} = (1 + \tfrac{1}{2}\mu\delta_y^2)V^n,$$

$$(1 - \tfrac{1}{2}\mu\delta_x^2)V^{n+1} = (1 + \tfrac{1}{2}\mu\delta_x^2)V^{n+\frac{1}{2}},$$

where $\mu = b\Delta t/(\Delta x)^2$. If b is a function of (x, y) and U^n denotes the solution of the ADI scheme, show that the schemes can still be related by setting $V^0 = (1 - \tfrac{1}{2}\mu\delta_y^2)U^0$, and find the relationship between U and V both at the end of a full time step, and at the intermediate stage.

4

Hyperbolic equations in one space dimension

4.1 Characteristics

The linear advection equation

$$\frac{\partial u}{\partial t} + a\frac{\partial u}{\partial x} = 0, \tag{4.1}$$

must surely be the simplest of all partial differential equations. Yet to approximate it well on a fixed (x,t)-mesh is a far from trivial problem that is still under active discussion in the numerical analysis literature. Of course, the exact solution is obtained from observing that this is a hyperbolic equation with a single set of characteristics and u is constant along each such characteristic: the *characteristics* are the solutions of the ordinary differential equation

$$\frac{\mathrm{d}x}{\mathrm{d}t} = a(x,t), \tag{4.2a}$$

and along a characteristic curve the solution $u(x,t)$ satisfies

$$\frac{\mathrm{d}u}{\mathrm{d}t} = \frac{\partial u}{\partial t} + \frac{\partial u}{\partial x}\frac{\mathrm{d}x}{\mathrm{d}t} = 0. \tag{4.2b}$$

Thus from initial data

$$u(x,0) = u^0(x), \tag{4.3}$$

where $u^0(x)$ is a given function, we can construct an approximate solution by choosing a suitable set of points $x_0, x_1, \ldots$, as in Fig. 4.1, and finding the characteristic through $(x_j, 0)$ by a numerical solution of (4.2a) with the initial condition $x(0) = x_j$. At all points on this curve we then have $u(x,t) = u^0(x_j)$. This is called the *method of characteristics*. Note that for this linear problem in which $a(x,t)$ is a given function, the characteristics cannot cross so long as a is Lipschitz continuous in x and continuous in t.

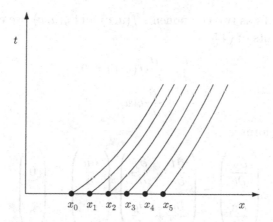

Fig. 4.1. Typical characteristics for $u_t + a(x,t)u_x = 0$.

When a is a constant the process is trivial. The characteristics are the parallel straight lines $x - at = $ constant, and the solution is simply

$$u(x,t) = u^0(x - at). \tag{4.4a}$$

Moreover, in the nonlinear problem in which a is a function only of u, $a = a(u)$, the characteristics are also straight lines because u is constant along each, although they are not now parallel. Thus again we are able to write the solution in the form

$$u(x,t) = u^0(x - a(u(x,t))t), \tag{4.4b}$$

until the time when this breaks down because the characteristics can now envelope or cross each other in some other manner – see Section 4.6.

Consideration of the characteristics of the equation, or system of equations, is essential in any development or study of numerical methods for hyperbolic equations and we shall continually refer to them below. We shall want to consider *systems of conservation laws* of the form

$$\frac{\partial \mathbf{u}}{\partial t} + \frac{\partial \mathbf{f}(\mathbf{u})}{\partial x} = \mathbf{0} \tag{4.5}$$

where $\mathbf{u} = \mathbf{u}(x,t)$ is a vector of unknown functions and $\mathbf{f}(\mathbf{u})$ a vector of *flux functions*. For example, if the vector $\mathbf{u}$ has two components

u and v, and $\mathbf{f}$ has two components $f(u, v)$ and $g(u, v)$, we can write out the components of (4.5) as

$$\frac{\partial u}{\partial t} + \frac{\partial}{\partial x} f(u, v) = 0, \tag{4.6a}$$

$$\frac{\partial v}{\partial t} + \frac{\partial}{\partial x} g(u, v) = 0, \tag{4.6b}$$

or in matrix form

$$\begin{pmatrix} \dfrac{\partial u}{\partial t} \\[2mm] \dfrac{\partial v}{\partial t} \end{pmatrix} + \begin{pmatrix} \dfrac{\partial f}{\partial u} & \dfrac{\partial f}{\partial v} \\[2mm] \dfrac{\partial g}{\partial u} & \dfrac{\partial g}{\partial v} \end{pmatrix} \begin{pmatrix} \dfrac{\partial u}{\partial x} \\[2mm] \dfrac{\partial v}{\partial x} \end{pmatrix} = \begin{pmatrix} 0 \\[2mm] 0 \end{pmatrix}. \tag{4.7}$$

If we define

$$A(\mathbf{u}) := \frac{\partial \mathbf{f}}{\partial \mathbf{u}}, \tag{4.8}$$

the Jacobian matrix formed from the partial derivatives of $\mathbf{f}$, we can write the system as

$$\mathbf{u}_t + A(\mathbf{u})\mathbf{u}_x = \mathbf{0}, \tag{4.9}$$

and the characteristic speeds are the eigenvalues of A. The hyperbolicity of the system is expressed by the fact that we assume A has real eigenvalues and a full set of eigenvectors. Suppose we denote by Λ the diagonal matrix of eigenvalues and by $S = S(\mathbf{u})$ the matrix of left eigenvectors, so that

$$SA = \Lambda S. \tag{4.10}$$

Then premultiplying (4.9) by S gives the *characteristic normal form* of the equations

$$S\mathbf{u}_t + \Lambda S\mathbf{u}_x = \mathbf{0}. \tag{4.11}$$

If it is possible to define a vector of *Riemann invariants* $\mathbf{r} = \mathbf{r}(\mathbf{u})$ such that $\mathbf{r}_t = S\mathbf{u}_t$ and $\mathbf{r}_x = S\mathbf{u}_x$, then we can write

$$\mathbf{r}_t + \Lambda \mathbf{r}_x = \mathbf{0} \tag{4.12}$$

which is a direct generalisation of the scalar case whose solution we have given in (4.4b). However, now each component of Λ will usually depend on all the components of $\mathbf{r}$ so that the characteristics will be curved. Moreover, although these Riemann invariants can always be defined for a system of two equations, for a larger system this is not always possible.

To apply the method of characteristics to problems like (4.5), where the characteristic speeds depend on the solution, one has to integrate forward simultaneously both the ordinary differential equations for the characteristic paths and the characteristic normal form (4.11) of the differential equations. This is clearly a fairly complicated undertaking, but it will give what is probably the most precise method for approximating this system of equations.

However, to use such a technique in a direct way in two space dimensions does become excessively complicated: for there we have characteristic surfaces and many more complicated solution phenomena to describe. Thus, even though this chapter is only on one-dimensional problems, in line with our general philosophy set out in the first chapter, we shall not consider the method of characteristics in any more detail. Instead we shall confine our considerations to methods based on a fixed mesh in space: and although the length of the time step may vary from step to step it must be the same over all the space points. We shall start with explicit methods on a uniform mesh.

4.2 The CFL condition

Courant, Friedrichs and Lewy, in their fundamental 1928 paper[1] on difference methods for partial differential equations, formulated a necessary condition now known as the *CFL condition* for the convergence of a difference approximation in terms of the concept of a domain of dependence. Consider first the simplest model problem (4.1), where a is a positive constant; as we have seen, the solution is $u(x,t) = u^0(x-at)$, where the function u^0 is determined by the initial conditions. The solution at the point (x_j, t_n) is obtained by drawing the characteristic through this point back to where it meets the initial line at $Q \equiv (x_j - at_n, 0)$ – see Fig. 4.2.

Now suppose that we compute a finite difference approximation by using the explicit scheme

$$\frac{U_j^{n+1} - U_j^n}{\Delta t} + a\frac{U_j^n - U_{j-1}^n}{\Delta x} = 0. \tag{4.13}$$

---

[1] Courant, R., Friedrichs, K.O. and Lewy, H. (1928), Über die partiellen Differenzengleichungen der mathematischen Physik, *Math. Ann.* **100**, 32–74.

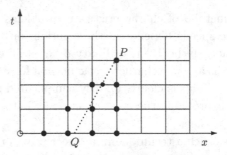

Fig. 4.2. Typical domain of dependence.

Then the value on the new time level will be calculated from

$$U_j^{n+1} = U_j^n - \frac{a\Delta t}{\Delta x}(U_j^n - U_{j-1}^n)$$
$$= (1 - \nu)U_j^n + \nu U_{j-1}^n, \qquad (4.14)$$

where

$$\nu = \frac{a\Delta t}{\Delta x}. \qquad (4.15)$$

The value of U_j^{n+1} depends on the values of U at two points on the previous time level; each of these depends on two points on the time level t_{n-1}, and so on. As illustrated in Fig. 4.2, the value of U_j^{n+1} depends on data given in a triangle with vertex (x_j, t_{n+1}), and ultimately on data at the points on the initial line

$$x_{j-n-1}, x_{j-n}, \ldots, x_{j-1}, x_j.$$

For an inhomogeneous equation in which a source term h_j^n replaces the zero on the right-hand side of (4.13), U_j^{n+1} depends on data given at all points of the triangle. This triangle is called the *domain of dependence* of U_j^{n+1}, or of the point (x_j, t_{n+1}), for this particular numerical scheme.

The corresponding domain of dependence of the differential equation is the characteristic path drawn back from (x_j, t_{n+1}) to the initial line, for in the inhomogeneous case $u_t + au_x = h$ data values $h(x, t)$ are picked up along the whole path as well as the initial data at $x = x_j - at_{n+1}$. The CFL condition then states that for a convergent scheme the domain of dependence of the partial differential equation must lie within the domain of dependence of the numerical scheme.

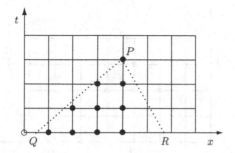

Fig. 4.3. Violation of the CFL condition.

Figure 4.3 illustrates two situations in which this condition is violated. Either of the characteristics PQ or PR lies outside the triangle. Suppose that we consider a refinement path on which the ratio $\Delta t/\Delta x$ is constant; then the triangular domain of dependence remains the same. But suppose we alter the given initial conditions in a small region of the initial line $t = 0$ around the point Q. This will then alter the solution of the differential equation at P, since the solution is constant along the characteristic QP. The numerical solution at P, however, remains unaltered, since the numerical data used to construct the solution are unchanged. The numerical solution therefore cannot converge to the required result at P. The same argument of course applies in the same way to the characteristic RP.

The CFL condition shows in this example that the scheme cannot converge for a differential equation for which $a < 0$, since this would give a characteristic like RP. And if $a > 0$ it gives a restriction on the size of the time step, for the condition that the characteristic must lie within the triangle of dependence requires that $a\Delta t/\Delta x \le 1$.

What we have thus obtained can also be regarded as a necessary condition for the stability of this difference scheme, somewhat similar to the condition for the stability of the explicit scheme for the parabolic equation in Chapter 2, but more obviously applicable to problems with variable coefficients, or even nonlinear problems. So far it is only a *necessary* condition. In general the CFL condition is not sufficient for stability, as we shall show in some examples. Its great merit lies in its simplicity; it enables us to reject a number of difference schemes with a trivial amount of investigation. Those schemes which satisfy the CFL condition may then be considered in more detail, using a test which is sufficient for stability.

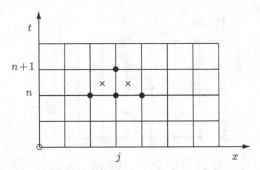

Fig. 4.4. General three-point scheme; the points marked ×
are used for the two-step Lax–Wendroff method.

Now suppose that we approximate the advection equation (4.1) by a
more general explicit scheme using just the three symmetrically placed
points at the old time level. The CFL condition becomes

$$|a|\Delta t \le \Delta x, \tag{4.16}$$

as we see from Fig. 4.4; $\nu := |a|\Delta t/\Delta x$ is often called the *CFL number*.

If $a > 0$, the difference scheme must use both U_{j-1}^n and U_j^n to obtain
U_j^{n+1}: and if $a < 0$ it must use U_j^n and U_{j+1}^n. To cover both cases we
might be tempted to use a central difference in space together with a
forward difference in time to obtain

$$\frac{U_j^{n+1} - U_j^n}{\Delta t} + a\frac{U_{j+1}^n - U_{j-1}^n}{2\Delta x} = 0. \tag{4.17}$$

If we satisfy (4.16) the CFL condition holds for either sign of a.

But now in the case where a is constant, and ignoring the effect of the
boundary conditions, we can investigate the stability of the scheme by
Fourier analysis, as we did for parabolic equations in Chapter 2. The
Fourier mode

$$U_j^n = (\lambda)^n e^{ik(j\Delta x)} \tag{4.18}$$

satisfies the difference scheme (4.17) provided that the amplification factor λ satisfies

$$\lambda \equiv \lambda(k) = 1 - (a\Delta t/\Delta x)i \sin k\Delta x. \tag{4.19}$$

Thus $|\lambda| > 1$ for all mesh ratios (and almost all modes) and the scheme
is unstable for any refinement path along which $a\Delta t/\Delta x$ is fixed. Note
that this is a case when the highest frequency mode, $k\Delta x = \pi$ or
$U_j \propto (-1)^j$, does not grow: but the mode with $k\Delta x = \frac{1}{2}\pi$, or where

U_j takes successive values $\ldots, -1, 0, 1, 0, -1, \ldots$, grows in magnitude by $[1 + (a\Delta t/\Delta x)^2]^{1/2}$ at each step while shifting to the right. This central difference scheme thus satisfies the CFL condition but is nevertheless always unstable, illustrating the earlier comment that the CFL condition is necessary, but not sufficient, for stability.

The simplest and most compact stable scheme involving these three points is called an *upwind scheme* because it uses a backward difference in space if a is positive and a forward difference if a is negative:

$$U_j^{n+1} = \begin{cases} U_j^n - a\dfrac{\Delta t}{\Delta x}\Delta_{+x}U_j^n & \text{if } a < 0, \\[2mm] U_j^n - a\dfrac{\Delta t}{\Delta x}\Delta_{-x}U_j^n & \text{if } a > 0. \end{cases} \tag{4.20}$$

If a is not a constant, but a function of x and t, we must specify which value is used in (4.20). We shall for the moment assume that we use $a(x_j, t_n)$, but still write a without superscript or subscript and $\nu = a\Delta t/\Delta x$ as in (4.15) when this is unambiguous.

This scheme clearly satisfies the CFL condition when (4.16) is satisfied, and a Fourier analysis gives for the constant $a > 0$ case the amplification factor

$$\lambda \equiv \lambda(k) = 1 - (a\Delta t/\Delta x)(1 - e^{-ik\Delta x}) \equiv 1 - \nu(1 - e^{-ik\Delta x}). \tag{4.21}$$

This leads to

$$\begin{aligned} |\lambda(k)|^2 &= [(1 - \nu) + \nu\cos k\Delta x]^2 + [\nu\sin k\Delta x]^2 \\ &= (1 - \nu)^2 + \nu^2 + 2\nu(1 - \nu)\cos k\Delta x \\ &= 1 - 2\nu(1 - \nu)(1 - \cos k\Delta x) \end{aligned}$$

which gives

$$|\lambda|^2 = 1 - 4\nu(1 - \nu)\sin^2 \tfrac{1}{2}k\Delta x. \tag{4.22}$$

It follows that $|\lambda(k)| \leq 1$ for all k provided that $0 \leq \nu \leq 1$. The same analysis for the case where $a < 0$ shows that the amplification factor $\lambda(k)$ is the same, but with a replaced by $|a|$. Thus in this case the CFL condition gives the correct stability limits, in agreement with the von Neumann condition as introduced in (2.56).

4.3 Error analysis of the upwind scheme

We notice that the scheme (4.20) can be written

$$U_j^{n+1} = \begin{cases} (1 + \nu)U_j^n - \nu U_{j+1}^n & \text{if } a < 0, \\ (1 - \nu)U_j^n + \nu U_{j-1}^n & \text{if } a > 0. \end{cases} \qquad (4.23)$$

This can be interpreted as follows. In Fig. 4.5 for the case $a > 0$, the characteristic through the point $P = (x_j, t_{n+1})$ meets the previous line $t = t_n$ at the point Q, which by the CFL condition must lie between the points $A = (x_j, t_n)$ and $B = (x_{j-1}, t_n)$. Moreover the exact solution $u(x,t)$ is constant along the characteristic, so that $u(P) = u(Q)$. Knowing an approximate numerical solution at all the points on the line t_n, we can therefore interpolate the value of $U(Q)$ and use this to give the required value U_j^{n+1}. If we use linear interpolation, approximating $u(x, t_n)$ by a linear function of x determined by the approximations at the two points A and B, we obtain (4.23) exactly when a is constant because $AQ = \nu \Delta x$ and $QB = (1 - \nu)\Delta x$; when a varies smoothly this still gives a good approximation.

Notice also that all the coefficients in (4.23) are nonnegative so that a maximum principle applies, provided that $|\nu| \leq 1$ at all mesh points. We can therefore obtain an error bound for the linear, variable coefficient problem just as we have done for parabolic equations. We must first consider more carefully what domain is given, and what conditions should be specified at the boundaries of the domain: although the physical problem may be given on the whole line, for all values of x, a numerical solution must be confined to a finite region. Suppose, for example, that the region of interest is $0 \leq x \leq X$, so that we have boundaries at $x = 0$ and $x = X$. Since the differential equation is hyperbolic and first order, we will usually have only one boundary condition; this is a fundamental difference from the parabolic equations of Chapter 2, where we were always given a boundary condition at each end of the domain. The direction of the characteristics shows that we need a boundary condition at $x = 0$ if $a > 0$ there, and at $x = X$ if $a < 0$ there; in the straightforward situation where a has the same sign everywhere, we therefore have just the one boundary condition. The exact solution of the differential equation would then be determined by drawing the characteristic backwards from the point P, until it reaches either the initial line $t = 0$, or a boundary on which a boundary condition is given.

For simplicity we shall first suppose that $a > 0$ on $[0, X] \times [0, t_F]$; we consider the general case later. The truncation error of the scheme is

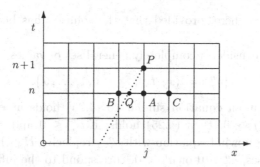

Fig. 4.5. Construction of a scheme by linear or quadratic interpolation.

defined as usual and expansion about (x_j, t_n) gives, if u is sufficiently smooth,

$$
\begin{aligned}
T_j^n &:= \frac{u_j^{n+1} - u_j^n}{\Delta t} + a_j^n \frac{u_j^n - u_{j-1}^n}{\Delta x} \\
&\sim [u_t + \tfrac{1}{2}\Delta t u_{tt} + \cdots]_j^n + [a(u_x - \tfrac{1}{2}\Delta x u_{xx} + \cdots))]_j^n \\
&= \tfrac{1}{2}(\Delta t u_{tt} - a\Delta x u_{xx}) + \cdots .
\end{aligned}
\tag{4.24}
$$

Even if a is constant so that we have $u_{tt} = a^2 u_{xx}$, we still find

$$
T_j^n = -\tfrac{1}{2}(1 - \nu)a\Delta x u_{xx} + \cdots ;
$$

hence generally the method is first order accurate. Suppose the difference scheme is applied for $j = 1, 2, \ldots, J$, at the points $x_j = j\Delta x$ with $J\Delta x = X$, and the boundary value $U_0^n = u(0, t_n)$ is given. Then for the error $e_j^n = U_j^n - u_j^n$ we have as usual

$$
e_j^{n+1} = (1 - \nu)e_j^n + \nu e_{j-1}^n - \Delta t T_j^n
\tag{4.25}
$$

and $e_0^n = 0$, from which we deduce that if $0 \le \nu \le 1$ at all points

$$
E^{n+1} := \max_j |e_j^{n+1}| \le E^n + \Delta t \max_j |T_j^n|.
$$

If we suppose that the truncation error is bounded, so that

$$
|T_j^n| \le T
\tag{4.26}
$$

for all j and n in the domain, the usual induction argument shows that

$$
E^n \le n\Delta t\, T \le t_F T
\tag{4.27}
$$

if $U_j^0 = u^0(x_j)$. This result is sufficient to prove first order convergence of the upwind scheme along a refinement path which satisfies the CFL condition everywhere, provided that the solution has bounded second derivatives.

Now let us consider a completely general set of values

$$\{a_j^n := a(x_j, t_n); \quad j = 0, 1, \ldots, J\}.$$

It is clear that an equation similar to (4.25) holds at each point: if $a_j^n \geq 0$ and $j > 0$, then (4.25) holds; if $a_j^n \leq 0$ and $j < J$ then a corresponding upwind equation with e_{j-1}^n replacing e_{j+1}^n holds; and the remaining cases, $a_0^n > 0$ or $a_J^n < 0$, correspond to the inflow boundary data being given so that either $e_0^{n+1} = 0$ or $e_J^{n+1} = 0$. The rest of the argument then follows as above.

We mentioned in Chapter 2 the difficulties which arise quite commonly in the analysis of parabolic equations when the given data function is discontinuous, or has discontinuous derivatives. In that case the solution itself had continuous derivatives in the interior of the region, and our difficulty lay in finding bounds on the derivatives. For hyperbolic equations the situation is quite different. We have seen that the solution of our model problem is constant along the characteristics. Suppose that the initial function $u(x, 0) = u^0(x)$ has a jump discontinuity in the first derivative at $x = \xi$. Then clearly the solution $u(x, t)$ also has a similar discontinuity at all points on the characteristic passing through the point $(\xi, 0)$; the discontinuity is not confined to the boundary of the domain. Such a solution satisfies the differential equation everywhere except along the line of the discontinuity, while satisfying (4.4b) everywhere; thus the latter can be regarded as defining a *generalised* solution of the differential equation, as distinct from a *classical solution* where the differential equation is satisfied at every point. Indeed, in this way we can define a solution for initial data $u^0(x)$ which itself has a jump discontinuity.

Most practical problems for hyperbolic systems involve discontinuities of some form in the given data, or arising in the solution; for such problems the analysis given above in terms of the global truncation error and the maximum norm is of little use, since the derivatives involved do not exist everywhere in the domain. For this reason, the truncation error is mainly of use for a local analysis, where in any case it would be better done by drawing the characteristic back to replace u_j^{n+1} by the value at its foot at time level t_n, i.e., $u(Q)$. The overall behaviour of a method, and its comparison with other methods, is often more satisfactorily carried out by means of Fourier analysis, an analysis of its conservation properties or a modified equation analysis – see later sections of this chapter and the next chapter.

4.4 Fourier analysis of the upwind scheme

Because hyperbolic equations often describe the motion and development of waves, Fourier analysis is of great value in studying the accuracy of methods as well as their stability. The modulus of $\lambda(k)$ describes the *damping* and the argument describes the *dispersion* in the scheme, i.e., the extent to which the wave speed varies with the frequency. We must, for the present and for a strict analysis, assume that a is a (positive) constant. The Fourier mode

$$u(x,t) = e^{i(kx+\omega t)} \tag{4.28}$$

is then an exact solution of the differential equation (4.1) provided that ω and k satisfy the *dispersion relation*

$$\omega = -ak. \tag{4.29}$$

The mode is completely undamped, as its amplitude is constant; in one time step its phase is changed by $-ak\Delta t$. By contrast, the Fourier mode (4.18) satisfies the upwind scheme provided that (4.21) holds. This leads to (4.22), showing that except in the special case $\nu = 1$ the mode is damped. The phase of the numerical mode is given by

$$\arg \lambda = -\tan^{-1}\left[\frac{\nu \sin k\Delta x}{(1-\nu) + \nu \cos k\Delta x}\right] \tag{4.30}$$

and we particularly need to evaluate this when $k\Delta x$ is small, as it is such modes that can be well approximated on the mesh. For this, and subsequent schemes, it is useful to have a simple lemma:

Lemma 4.1 *If q has an expansion in powers of p of the form*

$$q \sim c_1 p + c_2 p^2 + c_3 p^3 + c_4 p^4 + \cdots$$

as $p \to 0$, then

$$\tan^{-1} q \sim c_1 p + c_2 p^2 + (c_3 - \tfrac{1}{3}c_1^3)p^3 + (c_4 - c_1^2 c_2)p^4 + \cdots .$$

The proof of this lemma is left as an exercise.

We can now expand (4.30) and apply the lemma, giving

$$\begin{aligned}
\arg \lambda &\sim -\tan^{-1}\left[\nu(\xi - \tfrac{1}{6}\xi^3 + \cdots)(1 - \tfrac{1}{2}\nu\xi^2 + \cdots)^{-1}\right] \\
&= -\tan^{-1}\left[\nu\xi - \tfrac{1}{6}\nu(1 - 3\nu)\xi^3 + \cdots\right] \\
&= -\nu\xi[1 - \tfrac{1}{6}(1-\nu)(1-2\nu)\xi^2 + \cdots],
\end{aligned} \tag{4.31}$$

where we have written

$$\xi = k\Delta x. \tag{4.32}$$

The case $\nu = 1$ is obviously very special, as the scheme then gives the exact result. Apart from this, we have found that the upwind scheme always has an *amplitude error* which, from (4.22), is of order ξ^2 in one time step, corresponding to a global error of order ξ; and from (4.31) it has a *relative phase error* of order ξ^2, with the sign depending on the value of ν, and vanishing when $\nu = \frac{1}{2}$. The amplitude and relative phase errors are defined in more detail and plotted in Section 4.11, where they are compared with those of alternative schemes.

Some results obtained with the upwind scheme are displayed in Fig. 4.6. The problem consists of solving the equation

$$u_t + a(x,t)u_x = 0, \quad x \geq 0, \ t \geq 0, \tag{4.33a}$$

where

$$a(x,t) = \frac{1 + x^2}{1 + 2xt + 2x^2 + x^4}, \tag{4.33b}$$

with the initial condition

$$u(x,0) = \begin{cases} 1 & \text{if } 0.2 \leq x \leq 0.4, \\ 0 & \text{otherwise}, \end{cases} \tag{4.34a}$$

and the boundary condition

$$u(0,t) = 0. \tag{4.34b}$$

The exact solution of the problem is

$$u(x,t) = u(x^*,0) \tag{4.35a}$$

where

$$x^* = x - \frac{t}{1 + x^2}. \tag{4.35b}$$

Since $a(x,t) \leq 1$ the calculations use $\Delta t = \Delta x$, and the CFL stability condition is satisfied. The solution represents a square pulse moving to the right. It is clear from the figures how the damping of the high frequency modes has resulted in a substantial smoothing of the edges of the pulse, and a slight reduction of its height. However, the rather small phase error means that the pulse moves with nearly the right speed. The second set of results, with a halving of the mesh size in both co-ordinate directions, shows the expected improvement in accuracy, though the results are still not very satisfactory.

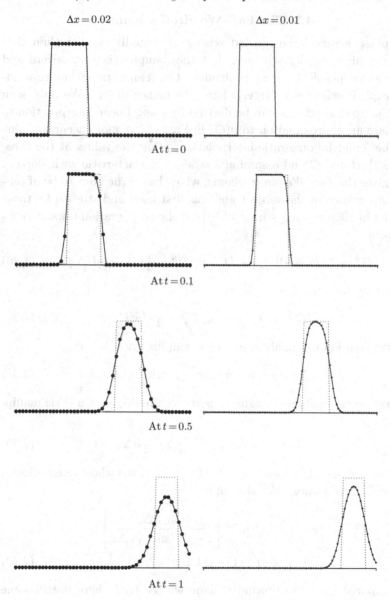

Fig. 4.6. Linear advection by the upwind method: problem (4.33), (4.34).

4.5 The Lax–Wendroff scheme

The phase error of the upwind scheme is actually smaller than that of many higher order schemes: but the damping is very severe and quite unacceptable in most problems. One can generate more accurate explicit schemes by interpolating to higher order. We have seen how the upwind scheme can be derived by using linear interpolation to calculate an approximation to $u(Q)$ in Fig. 4.5. A more accurate value may be found by quadratic interpolation, using the values at the three points A, B and C and assuming a straight characteristic with slope ν. This gives the *Lax–Wendroff* scheme, which has turned out to be of central importance in the subject and was first used and studied by those authors in 1960 in their study[1] of hyperbolic conservation laws; it takes the form

$$U_j^{n+1} = \tfrac{1}{2}\nu(1+\nu)U_{j-1}^n + (1-\nu^2)U_j^n - \tfrac{1}{2}\nu(1-\nu)U_{j+1}^n \qquad (4.36)$$

which may be written

$$U_j^{n+1} = U_j^n - \nu\Delta_{0x}U_j^n + \tfrac{1}{2}\nu^2\delta_x^2 U_j^n. \qquad (4.37)$$

The usual Fourier analysis gives the amplification factor

$$\lambda(k) = 1 - i\nu\sin k\Delta x - 2\nu^2\sin^2\tfrac{1}{2}k\Delta x. \qquad (4.38)$$

Separating the real and imaginary parts we obtain, after a little manipulation,

$$|\lambda|^2 = 1 - 4\nu^2(1-\nu^2)\sin^4\tfrac{1}{2}k\Delta x. \qquad (4.39)$$

Thus we see that the scheme is stable for $|\nu| \leq 1$, the whole range allowed by the CFL condition. We also find

$$\arg\lambda = -\tan^{-1}\left[\frac{\nu\sin k\Delta x}{1 - 2\nu^2\sin^2\tfrac{1}{2}k\Delta x}\right]$$

$$\sim -\nu\xi\left[1 - \tfrac{1}{6}(1-\nu^2)\xi^2 + \cdots\right]. \qquad (4.40)$$

Compared with the upwind scheme we see that there is still some damping, as in general $|\lambda| < 1$, but the amplitude error in one time step is now of order ξ^4 when ξ is small, compared with order ξ^2 for the upwind scheme; this is a substantial improvement. Both the schemes have a relative phase error of order ξ^2, which are equal when $\nu \sim 0$; but

---

[1] Lax, P.D. and Wendroff, B. (1960), Systems of conservation laws, *Comm. Pure and Appl. Math.* **13**, 217–37.

the error is always of one sign (corresponding to a phase lag) for Lax–Wendroff while it goes through a zero at $\nu = \frac{1}{2}$ for the upwind scheme. However, the much smaller damping of the Lax–Wendroff scheme often outweighs the disadvantage of the larger phase error.

In deriving the Lax–Wendroff scheme above we assumed a was constant. To deal with variable a in the linear equation (4.1) we derive it in a different way, following the original derivation. We first expand in a Taylor series in the variable t, giving

$$u(x, t + \Delta t) = u(x, t) + \Delta t \, u_t(x, t) + \tfrac{1}{2}(\Delta t)^2 u_{tt}(x, t) + O((\Delta t)^3). \quad (4.41)$$

Then we convert the t-derivatives into x-derivatives by using the differential equation, so that

$$u_t = -au_x, \quad (4.42a)$$

$$u_{tt} = -a_t u_x - au_{xt}, \quad (4.42b)$$

$$u_{xt} = u_{tx} = -(au_x)_x, \quad (4.42c)$$

which give

$$u_{tt} = -a_t u_x + a(au_x)_x. \quad (4.43)$$

Approximating each of these x-derivatives by central differences gives the scheme

$$U_j^{n+1} = U_j^n - a_j^n \Delta t \frac{\Delta_{0x} U_j^n}{\Delta x}$$
$$+ \tfrac{1}{2}(\Delta t)^2 \left[-(a_t)_j^n \frac{\Delta_{0x} U_j^n}{\Delta x} + a_j^n \frac{\delta_x(a_j^n \delta_x U_j^n)}{(\Delta x)^2} \right]. \quad (4.44)$$

This scheme involves evaluating the function $a(x, t)$ at the points $x = x_j \pm \frac{1}{2}\Delta x$ as well as a and a_t at x_j. Note, however, that the scheme can be simplified by replacing $a_j^n + \frac{1}{2}\Delta t(a_t)_j^n$ by $a_j^{n+1/2}$ in the coefficient of $\Delta_{0x} U_j^n$; see also the next section for conservation laws with $au_x \equiv f_x$, and also the following section on finite volume schemes.

The results in Fig. 4.7 are obtained by applying this scheme to the same problem (4.33), (4.34) used to test the upwind scheme, with the same mesh sizes. Comparing the results of Fig. 4.6 and Fig. 4.7 we see that the Lax–Wendroff scheme maintains the height and width of the pulse rather better than the upwind scheme, which spreads it out much more. On the other hand, the Lax–Wendroff scheme produces oscillations which follow behind the two discontinuities as the pulse moves to the right. Notice also that the reduction in the mesh size Δx does

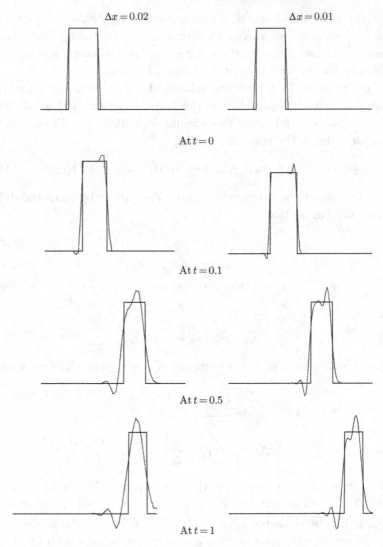

Fig. 4.7. Linear advection by the Lax–Wendroff method: problem (4.33), (4.34).

improve the accuracy of the result, but not by anything like the factor of 4 which would be expected of a scheme for which the error is $O((\Delta x)^2)$. The analysis of truncation error is only valid for solutions which are sufficiently smooth, while this problem has a discontinuous solution. In fact the maximum error in this problem is $O((\Delta x)^{1/2})$ for the upwind scheme and $O((\Delta x)^{2/3})$ for the Lax–Wendroff scheme. The error therefore tends to zero rather slowly as the mesh size is reduced.

The oscillations in Fig. 4.7 arise because the Lax–Wendroff scheme does not satisfy a maximum principle. We see from (4.36) that with $\nu > 0$ the coefficient of U_{j+1}^n is negative, since we require that $\nu \leq 1$ for stability. Hence U_j^{n+1} is given as a weighted mean of three values on the previous time level, but two of the weights are positive and one is negative. It is therefore possible for the numerical solution to have oscillations with internal maxima and minima.

As an example of a problem with a smooth solution, we consider the same equation as before, (4.33a,b), but replace the initial condition (4.34) by

$$u(x, 0) = \exp[-10(4x - 1)^2].\tag{4.45}$$

The results are illustrated in Fig. 4.8. As before, the solution consists of a pulse moving to the right, but now the pulse has a smooth Gaussian shape, instead of a discontinuous square wave. Using the same mesh sizes as before, the results are considerably more accurate. There is still some sign of an oscillation to the left of the pulse by the time that $t = 1$, but it is a good deal smaller than in the discontinuous case. Moreover, the use of the smaller mesh size has reduced the size of the errors and this oscillation becomes nearly invisible.

4.6 The Lax–Wendroff method for conservation laws

In practical situations a hyperbolic equation often appears in the form

$$\frac{\partial u}{\partial t} + \frac{\partial f(u)}{\partial x} = 0\tag{4.46}$$

which may be written in the form we have considered above,

$$u_t + au_x = 0,\tag{4.47}$$

where $a = a(u) = \partial f / \partial u$. It is then convenient to derive the Lax–Wendroff scheme directly for the conservation form (4.46). The function f does not involve x or t explicitly but is a function of u only. The t-derivatives required in the Taylor series expansion (4.41) can now be written

$$u_t = -(f(u))_x\tag{4.48a}$$

and

$$u_{tt} = -f_{xt} = -f_{tx} = -(au_t)_x = (af_x)_x.\tag{4.48b}$$

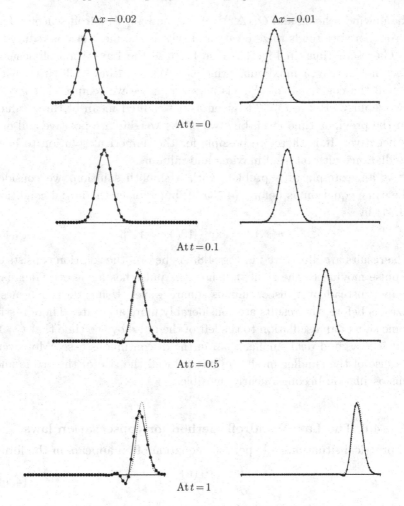

Fig. 4.8. Linear advection by the Lax–Wendroff method: (4.33) with the data (4.45).

Replacing the x-derivatives by central differences as before we now obtain

$$U_j^{n+1} = U_j^n - \frac{\Delta t}{\Delta x} \Delta_{0x} f(U_j^n)$$

$$+ \tfrac{1}{2} \left(\frac{\Delta t}{\Delta x} \right)^2 \delta_x \left[a(U_j^n) \delta_x f(U_j^n) \right]. \tag{4.49}$$

It is clear that this reduces to (4.37) when $f(u) = au$ where a is constant. If we expand the last term in (4.49) we see that it involves the values of $a(U_{j-1/2}^n)$ and $a(U_{j+1/2}^n)$; in evaluating these we could set $U_{j\pm1/2}^n :=$ $\frac{1}{2}(U_j^n + U_{j\pm1}^n)$, but a commonly used alternative is to replace them by $\Delta_{\pm x}f(U_j^n)/\Delta_{\pm x}U_j^n$. Then writing F_j^n for $f(U_j^n)$ and $A_{j\pm1/2}^n$ for either choice of the characteristic speeds, the scheme becomes

$$
U_j^{n+1} = U_j^n - \frac{1}{2}\frac{\Delta t}{\Delta x}\left\{\left[1 - A_{j+1/2}^n\frac{\Delta t}{\Delta x}\right]\Delta_{+x}F_j^n\right.
$$
$$
\left. + \left[1 + A_{j-1/2}^n\frac{\Delta t}{\Delta x}\right]\Delta_{-x}F_j^n\right\}. \qquad (4.50)
$$

As an example of the use of this scheme we consider the limiting case of Burgers' equation, for inviscid flow,

$$
u_t + u\,u_x = 0, \qquad (4.51)
$$

or in conservation form

$$
u_t + (\tfrac{1}{2}u^2)_x = 0. \qquad (4.52)
$$

The general solution when it is smooth is easily obtained by the method of characteristics, or it is sufficient to verify that the solution is given implicitly by

$$
u \equiv u(x,t) = u^0(x - tu(x,t)). \qquad (4.53)
$$

The characteristics are straight lines, and the solution $u(x,t)$ is constant along each of them. Given the initial condition $u(x,0) = u^0(x)$, they are obtained by drawing the straight line with slope $dt/dx = 1/u^0(x_0)$ through the point $(x_0,0)$, for each value of x_0.

If we use the smooth initial data given by (4.45), we find a typical example of a common difficulty in nonlinear hyperbolic equations. Since the solution is constant along each characteristic, a singularity will arise wherever two characteristics cross; they must have different slopes at a point where they cross, and so the solution will have different values on the two characteristics, and must become multivalued. This situation must arise whenever the initial function $u^0(x)$ has a decreasing derivative. Some typical characteristics are illustrated in Fig. 4.9; in problems like this there will be a critical value t_c such that the solution exists and is single-valued for $0 \le t < t_c$, but a singularity appears at $t = t_c$.

This behaviour is a simple model for the formation of *shocks* in the flow of a gas. Not only does the classical solution of (4.51) break down when the characteristics cross, but so too does the mathematical model of the

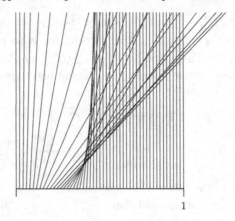

Fig. 4.9. Typical characteristics for the inviscid Burgers'
equation.

physical situation. Viscosity becomes important in the steep gradients
that occur and the full, viscous, Burgers' equation $u_t + uu_x = \nu_v u_{xx}$
should be used. A thorough description of the situation is beyond the
scope of this book but some key points are very pertinent to our emphasis
on the use of the conservation law forms of equations.

What one can hope to approximate beyond the point of breakdown is
the limit of the solution to the viscous equation as the viscosity ν_v tends
to zero. This will have a discontinuity, representing a shock, which for
the conservation law $u_t + f_x = 0$ will move at the *shock speed*

$$S := \frac{[f(u)]}{[u]}, \qquad (4.54)$$

where $[u]$ denotes the jump in the variable u; thus if the limiting value
of u on the left is u_L and on the right is u_R, the shock moves at a speed

$$\frac{f(u_R) - f(u_L)}{u_R - u_L},$$

which clearly tends to $a(u_L)$ as $u_R \to u_L$, i.e., in the limit of a 'weak'
shock. Such a relation can be deduced by integrating the equation over
a small box in the (x, t)-plane, aligned with and covering a portion of
the shock, and then applying the Gauss divergence theorem. Solutions
of the differential equation which are only satisfied in this averaged way
are called *weak solutions*.

Note, however, that the speed obtained in this way depends on the form of the conservation law and, for example, from the non-conservative form (4.51) we could deduce any law of the form

$$((m+1)u^m)_t + (mu^{m+1})_x = 0. \tag{4.55}$$

Only one choice correctly models the more complicated physical situation. In the present case $m = 1$ is correct, namely (4.52), and thus this is not equivalent to, but is more general than, (4.51). A shock from u_L to u_R therefore moves at the mean speed $\frac{1}{2}(u_L + u_R)$, and not for instance at $\frac{2}{3}(u_L^2 + u_L u_R + u_R^2)/(u_L + u_R)$ which it would if we had $m = 2$.

In Fig. 4.10 are shown early results obtained from the conservation law (4.52) using the initial data (4.45). A shock develops at $t_c = 0.25$ approximately and grows in strength, though the whole solution will eventually decay to zero. This weak solution of the equation is shown as a dotted line, with its approximation by the Lax–Wendroff scheme shown on the left. For comparison we have also shown on the right of Fig. 4.10 the approximation obtained with the upwind scheme, which we write in the form

$$U_j^{n+1} = U_j^n - \frac{1}{2}\frac{\Delta t}{\Delta x}\Big\{ \Big[1 - \operatorname{sgn} A_{j+\frac{1}{2}}^n\Big]\Delta_{+x}F_j^n$$
$$+ \Big[1 + \operatorname{sgn} A_{j-\frac{1}{2}}^n\Big]\Delta_{-x}F_j^n\Big\} \tag{4.56}$$

where the preferred choice is $A_{j\pm\frac{1}{2}}^n := \Delta_{\pm x}F_j^n/\Delta_{\pm x}U_j^n$, reducing to $a(U_j^n)$ when $U_j^n = U_{j\pm1}^n$; this form clearly generalises (4.20) and is directly comparable with (4.50). The greater accuracy of the Lax–Wendroff method away from the shock is apparent from these figures; but the oscillations that develop behind the shock, in contrast to their absence with the upwind method, prompt the idea of adaptively switching between the two schemes, as pioneered in the work of van Leer.[1]

One of the great strengths of the Lax–Wendroff method is that it can be extended quite easily to systems of equations. Instead of (4.46) and (4.48a,b) we have

$$\mathbf{u}_t = -\mathbf{f}_x, \quad \mathbf{u}_{tt} = -\mathbf{f}_{tx} = -(A\mathbf{u}_t)_x = (A\mathbf{f}_x)_x \tag{4.57}$$

---

[1] van Leer, B. (1974), Towards the ultimate conservative difference scheme. II monotonicity and conservation combined in a second order scheme, *J. of Comput. Phys.* **14**, 361–70.

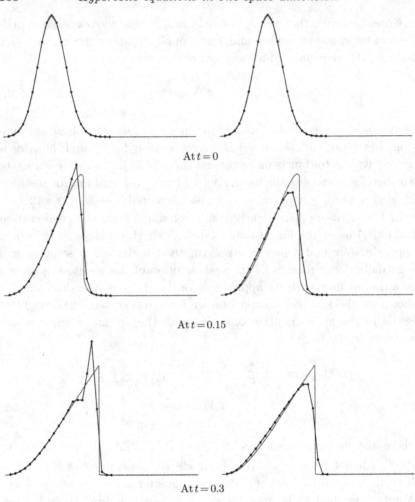

Fig. 4.10. Burgers' equation with initial data (4.45), approximated by the Lax–Wendroff method (on the left) and the
upwind method (on the right).

where A is the Jacobian matrix as in (4.8); then (4.49) simply becomes

$$\mathbf{U}_j^{n+1} = \mathbf{U}_j^n - \left(\frac{\Delta t}{\Delta x}\right) \Delta_{0x} \mathbf{f}(\mathbf{U}_j^n)$$

$$+ \tfrac{1}{2} \left(\frac{\Delta t}{\Delta x}\right)^2 \delta_x \left[A(\mathbf{U}_j^n) \delta_x \mathbf{f}(\mathbf{U}_j^n) \right], \qquad (4.58)$$

and (4.50) is exactly the same except for the use of vectors $\mathbf{U}_j^n$ and $\mathbf{F}_j^n$.

In the special case where $\mathbf{f}(\mathbf{u}) = A\mathbf{u}$ and A is a constant matrix, this reduces to

$$\mathbf{U}_j^{n+1} = \mathbf{U}_j^n - \left(\frac{\Delta t}{\Delta x}\right) A\Delta_{0x}\mathbf{U}_j^n + \tfrac{1}{2}\left(\frac{\Delta t}{\Delta x}\right)^2 A^2\delta_x^2\mathbf{U}_j^n. \qquad (4.59)$$

For this problem a Fourier analysis is possible. Each of the components of the vector $\mathbf{U}$ will be a multiple of the same Fourier mode, and we look for a solution of (4.59) which is of the form

$$\mathbf{U}_j^n = \lambda^n e^{ikj\Delta x}\widehat{\mathbf{U}} \qquad (4.60)$$

where $\widehat{\mathbf{U}}$ is a constant vector. This is a solution provided that

$$\left\{\lambda I - \left[I - i\left(\frac{\Delta t}{\Delta x}\right)\sin k\Delta x A - 2\left(\frac{\Delta t}{\Delta x}\right)^2 \sin^2 \tfrac{1}{2}k\Delta x A^2\right]\right\}\widehat{\mathbf{U}} = \mathbf{0}.$$
$$\qquad (4.61)$$

For this to hold $\widehat{\mathbf{U}}$ has to be an eigenvector of A; if μ is the corresponding eigenvalue we write $\nu = \mu\Delta t/\Delta x$ and obtain

$$\lambda = 1 - i\nu\sin k\Delta x - 2\nu^2\sin^2 \tfrac{1}{2}k\Delta x \qquad (4.62)$$

which is precisely the same as (4.38). Thus we can deduce a necessary condition for the scheme to be stable as

$$\frac{\rho\Delta t}{\Delta x} \leq 1 \qquad (4.63)$$

where ρ is the *spectral radius* of A, the largest of the magnitudes of the eigenvalues of A, which is a generalisation of our earlier concept of stability to systems of equations. We leave until the next chapter a consideration of whether this necessary von Neumann condition is a sufficient condition for stability.

It is an important advantage of the Lax–Wendroff scheme that the stability condition involves only the magnitudes of the eigenvalues, not their signs, so that its form does not have to be changed with a switch in sign. We have seen in (4.20) and (4.56) how the upwind scheme for a single equation uses either a forward or backward difference, according to the sign of a. This is much more difficult for a system of equations. It requires that, at each point, we find the eigenvalues and eigenvectors of an approximate Jacobian matrix $\tilde{A}$, express the current vector in terms of the eigenvectors, and use forward or backward differences for each eigenvector according to the sign of the corresponding eigenvalue; the eigenvectors are then combined together again to give the solution at the new time level. The Lax–Wendroff method avoids this considerable complication, but at the cost of the oscillations shown in Figs. 4.7 and 4.10. There has thus been considerable development of methods which combine the advantages of the two approaches, as has already been mentioned and as we will discuss further in the next section. Although well beyond the scope of this book, it is worth noting here that the generalisation of the shock relation (4.54) – and of the characteristic speeds $A^n_{j\pm1/2}$ used in the upwind scheme (4.56) – which the approximate Jacobian $\tilde{A}$ is usually made to satisfy, is given by

$$\tilde{A}^n_{j\pm1/2}\Delta_{\pm x}\mathbf{U}^n_j = \Delta_{\pm x}\mathbf{F}^n_j. \qquad (4.64)$$

Following the work of Roe[1] for the gas dynamics equations, this is an important relation in the development of many modern methods for conservation laws.

Lastly, a convenient and often-used variant of the Lax–Wendroff scheme is the two-step method. Values are predicted at points $(x_{j+1/2}, t_{n+1/2})$ and then a centred scheme used to obtain the final values at t_{n+1} (see Fig. 4.4):

$$\mathbf{U}^{n+1/2}_{j+1/2} = \tfrac{1}{2}\left(\mathbf{U}^n_j + \mathbf{U}^n_{j+1}\right)$$
$$-\tfrac{1}{2}(\Delta t/\Delta x)\left[\mathbf{f}\left(\mathbf{U}^n_{j+1}\right) - \mathbf{f}\left(\mathbf{U}^n_j\right)\right], \qquad (4.65a)$$
$$\mathbf{U}^{n+1}_j = \mathbf{U}^n_j - (\Delta t/\Delta x)\left[\mathbf{f}\left(\mathbf{U}^{n+1/2}_{j+1/2}\right) - \mathbf{f}\left(\mathbf{U}^{n+1/2}_{j-1/2}\right)\right]. \qquad (4.65b)$$

For the linear case where $\mathbf{f} = A\mathbf{u}$ with constant A, we leave it as an exercise to show that on elimination of the intermediate value $\mathbf{U}^{n+1/2}_{j+1/2}$ we obtain exactly the same result as the standard one-step Lax–Wendroff method (4.59). For nonlinear problems and variable coefficients the two variants do not yield the same results. One of the most important advantages of (4.65) is that it avoids the need to calculate the Jacobian matrix A.

4.7 Finite volume schemes

Many of the methods that are used for practical computation with conservation laws are classed as *finite volume methods*, and that in (4.65) is a typical example. Suppose we take the system of equations $\mathbf{u}_t + \mathbf{f}_x = 0$

[1] Roe, P.L. (1981), Approximate Riemann solvers, parameter vectors, and difference schemes, *J. of Comput. Phys.* **43**, 357–72.

in conservation law form and integrate over a region Ω in (x, t)-space; using the Gauss divergence theorem this becomes a line integral,

$$\iint_\Omega (\mathbf{u}_t + \mathbf{f}_x)\,\mathrm{d}x\,\mathrm{d}t \equiv \iint_\Omega \mathrm{div}(\mathbf{f}, \mathbf{u})\,\mathrm{d}x\,\mathrm{d}t$$

$$= \oint_{\partial\Omega} [\mathbf{f}\,\mathrm{d}t - \mathbf{u}\,\mathrm{d}x]. \tag{4.66}$$

In particular, if we take the region to be a rectangle of width Δx and height Δt and introduce averages along the sides, such as $\mathbf{u}_{\mathrm{top}}$ etc., we obtain

$$(\mathbf{u}_{\mathrm{top}} - \mathbf{u}_{\mathrm{bottom}})\Delta x + (\mathbf{f}_{\mathrm{right}} - \mathbf{f}_{\mathrm{left}})\Delta t = 0. \tag{4.67}$$

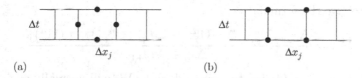

(a)   (b)

Fig. 4.11. Two finite volume schemes: (a) with mid-point quadrature; (b) with trapezoidal quadrature.

Then to obtain a specific numerical scheme these averages need to be approximated by some form of quadrature. For instance, we can use mid-point quadrature on all four sides – see Fig. 4.11(a): if we denote by $\mathbf{U}_j^n$ the approximate solution at time level n at the centre of cell j of width Δx_j, and by $\mathbf{F}_{j+1/2}^{n+1/2}$ the flux value halfway up a cell side, we obtain the scheme

$$\mathbf{U}_j^{n+1} = \mathbf{U}_j^n - (\Delta t/\Delta x_j)\left(\mathbf{F}_{j+1/2}^{n+1/2} - \mathbf{F}_{j-1/2}^{n+1/2}\right). \tag{4.68}$$

It remains to calculate the fluxes from the set of $\mathbf{U}_j^n$ values, for example by the Taylor expansion used in the two-step Lax–Wendroff method: that is, solution values on the cell sides are calculated by the formula (4.65a) and these are substituted into equation (4.65b), which is exactly of the form (4.68).

Note, however, that in (4.68) we have allowed for the cell widths to be quite arbitrary. This is a great advantage of this formulation, and is very useful in practical calculations – even more so in more space dimensions. Thus, for instance, we can sum the integrals over a set of contiguous cells to obtain from (4.68)

$$\sum_{j=k}^{l} \Delta x_j \left(\mathbf{U}_j^{n+1} - \mathbf{U}_j^n \right) + \Delta t \left(\mathbf{F}_{l+1/2}^{n+1/2} - \mathbf{F}_{k-1/2}^{n+1/2} \right) = 0, \qquad (4.69)$$

which exactly mirrors the conservation property of the differential equation. In the case of the Lax–Wendroff scheme, though, if $\mathbf{U}_j^n$ is taken to represent the solution at the cell centre then we need to use a Taylor expansion at a cell edge $x_{j+1/2}$ to give, to the required first order accuracy,

$$\mathbf{u}(x_{j+1/2}, t_n + \Delta t/2) = \mathbf{u}(x_{j+1/2}, t_n)$$
$$- \tfrac{1}{2}\Delta t \mathbf{f}_x(x_{j+1/2}, t_n) + O\big((\Delta t)^2\big);$$

this can be combined with expansions for the cell centre values on either side to give the formula

$$\mathbf{U}_{j+1/2}^{n+1/2} = \frac{\Delta x_{j+1} \mathbf{U}_j^n + \Delta x_j \mathbf{U}_{j+1}^n - \Delta t \left[\mathbf{f}\left(\mathbf{U}_{j+1}^n \right) - \mathbf{f}\left(\mathbf{U}_j^n \right) \right]}{\Delta x_j + \Delta x_{j+1}} \qquad (4.70)$$

which generalises (4.65a) to a general mesh. We will normally avoid this extra complexity in what follows and revert to our usual assumption of uniform mesh spacing.

As we have already noted and demonstrated, a major disadvantage of the Lax–Wendroff method is its proneness to produce oscillatory solutions. The problem has prompted much of the development of finite volume methods, and can be fully analysed for scalar conservation laws. The guiding principle is provided by controlling the *total variation* of the solution: on a finite domain $[0, X]$ divided into J cells, with U^n taking the value U_j^n in cell j at time level n, we can define the total variation as

$$\mathrm{TV}(U^n) := \sum_{j=1}^{J-1} \left| U_{j+1}^n - U_j^n \right| \equiv \sum_{j=1}^{J-1} \left| \Delta_{+x} U_j^n \right|. \qquad (4.71)$$

More generally, for the exact solution $u(x, t)$, $\mathrm{TV}(u(\cdot, t))$ can be defined by taking the supremum, over all subdivisions of the $[0, X]$ interval such as $0 = \xi_0 < \xi_1 < \cdots < \xi_K = X$, of the sum of the corresponding differences $|u(\xi_{j+1}, t) - u(\xi_j, t)|$. Clearly, these are consistent definitions when U^n is regarded as a piecewise constant approximation to $u(\cdot, t_n)$. To simplify the subsequent discussion, however, by leaving aside the specification of boundary conditions, we will assume that both $u(\cdot, t)$

and U^n are extended by constant values to the left and right so that the range of the summation over j will not be specified.

A key property of the solution of a conservation law such as (4.46) is that $\mathrm{TV}(u(\cdot, t))$ is a nonincreasing function of t – which can be deduced informally from the constancy of the solution along the characteristics described by (4.4b). Thus we define TVD(*total variation diminishing*) schemes as those for which we have $\mathrm{TV}(U^{n+1}) \leq \mathrm{TV}(U^n)$. This concept is due to Harten[1] who established the following useful result:

Theorem 4.1 (Harten) *A scheme is TVD if it can be written in the form*

$$U_j^{n+1} = U_j^n - C_{j-1}\Delta_{-x}U_j^n + D_j\Delta_{+x}U_j^n, \tag{4.72}$$

where the coefficients C_j and D_j, which may be any functions of the solution variables $\{U_j^n\}$, satisfy the conditions

$$C_j \geq 0, D_j \geq 0 \text{ and } C_j + D_j \leq 1 \ \forall j. \tag{4.73}$$

Proof Taking the forward difference of (4.72), and freely using the identity $\Delta_{+x}U_j \equiv \Delta_{-x}U_{j+1}$, we get

$$\begin{aligned}
U_{j+1}^{n+1} - U_j^{n+1} &= \Delta_{+x}U_j^n - C_j\Delta_{+x}U_j^n + C_{j-1}\Delta_{-x}U_j^n \\
&\quad + D_{j+1}\Delta_{+x}U_{j+1}^n - D_j\Delta_{+x}U_j^n \\
&= (1 - C_j - D_j)\Delta_{+x}U_j^n + C_{j-1}\Delta_{-x}U_j^n \\
&\quad + D_{j+1}\Delta_{+x}U_{j+1}^n.
\end{aligned}$$

By the hypotheses of (4.73), all the coefficients on the right of this last expression are nonnegative. So we can take absolute values to obtain

$$\begin{aligned}
\left|\Delta_{+x}U_j^{n+1}\right| &\leq (1 - C_j - D_j)\left|\Delta_{+x}U_j^n\right| \\
&\quad + C_{j-1}\left|\Delta_{-x}U_j^n\right| + D_{j+1}\left|\Delta_{+x}U_{j+1}^n\right|,
\end{aligned}$$

then summing over j leads to cancellation and hence the result $\mathrm{TV}(U^{n+1}) \leq \mathrm{TV}(U^n)$. $\qquad\square$

Suppose we attempt to apply this theorem to both the Lax–Wendroff method and the upwind method. We consider the latter first, in the form given in (4.56) with $A_{j\pm1/2}^n := \Delta_{\pm x}F_j^n/\Delta_{\pm x}U_j^n$. This corresponds to the scalar case of the scheme due to Roe, referred to in the sentence

[1] Harten, A. (1984), On a class of high resolution total-variation-stable finite-difference schemes, *SIAM J. Numer. Anal.* **21**, 1–23.

following (4.64), and is best considered as a finite volume scheme in which the fluxes of (4.68) are given by

$$F_{j+1/2}^{n+1/2} = \begin{cases} f(U_j^n) & \text{when} \quad A_{j+1/2}^n \geq 0, \\ f(U_{j+1}^n) & \text{when} \quad A_{j+1/2}^n < 0; \end{cases} \qquad (4.74)$$

or, equivalently,

$$F_{j+1/2}^{n+1/2} = \tfrac{1}{2} \left[\left(1 + \operatorname{sgn} A_{j+1/2}^n \right) F_j^n + \left(1 - \operatorname{sgn} A_{j+1/2}^n \right) F_{j+1}^n \right]. \qquad (4.75)$$

Then, comparing (4.56) with (4.72) after replacing the flux difference $\Delta_{-x} F_j^n$ by $A_{j-1/2}^n \Delta_{-x} U_j^n$, we are led to setting

$$C_{j-1} = \tfrac{1}{2} \frac{\Delta t}{\Delta x} \left(1 + \operatorname{sgn} A_{j-1/2}^n \right) A_{j-1/2}^n.$$

This is clearly always nonnegative, thus satisfying the first condition of (4.73). Similarly, we set

$$D_j = -\tfrac{1}{2} \frac{\Delta t}{\Delta x} \left(1 - \operatorname{sgn} A_{j+1/2}^n \right) A_{j+1/2}^n,$$

which is also nonnegative. Moreover, adding the two together and remembering the shift of subscript in the former, we get

$$C_j + D_j = \tfrac{1}{2} \frac{\Delta t}{\Delta x} \left[\left(1 + \operatorname{sgn} A_{j+1/2}^n \right) A_{j+1/2}^n + \left(1 - \operatorname{sgn} A_{j+1/2}^n \right) A_{j+1/2}^n \right]$$
$$\equiv \left| A_{j+1/2}^n \right| \frac{\Delta t}{\Delta x},$$

which is just the CFL number. Hence the last condition of (4.73) corresponds to the CFL stability condition; we have shown that the Roe first order upwind scheme is TVD when Δt is chosen so that it is stable.

On the other hand, if we attempt to follow similar arguments with the Lax–Wendroff scheme in the corresponding form of (4.50) and write $\nu_{j\pm1/2}^n$ for $A_{j\pm1/2}^n \Delta t / \Delta x$, we are led to setting

$$C_j = \tfrac{1}{2} \nu_{j+1/2}^n (1 + \nu_{j+1/2}^n), \quad \text{and} \quad D_j = -\tfrac{1}{2} \nu_{j+1/2}^n (1 - \nu_{j+1/2}^n), \qquad (4.76)$$

both of which have to be nonnegative. Then the third condition of (4.73) requires that the CFL condition $(\nu_{j+1/2}^n)^2 \leq 1$ be satisfied, and the only values that $\nu_{j+1/2}^n$ can take to satisfy all three conditions are $-1, 0$ and $+1$; this is clearly impractical for anything other than very special cases.

The TVD property of the Roe upwind scheme has made it a very important building block in the development of more sophisticated finite volume methods, and it is particularly successful in modelling shocks. However, it needs modification for the handling of some rarefaction waves. For instance, suppose that the inviscid Burgers equation (4.52) is given the initial data $\{U_j^0 = -1, \text{ for } j \leq 0; U_j^0 = +1, \text{ for } j > 0\}$, which should lead to a spreading rarefaction wave. Then it is clear from (4.74) that in Roe's scheme all the fluxes would be equal to $\frac{1}{2}$ so that the solution would not develop at all. The problem is associated with the *sonic points* which occurs for $u = 0$ where the characteristic speed a is zero – more precisely, with the *transonic rarefaction wave* that we have here in which the characteristic speeds are negative to the left of this point and positive to the right. For a general convex flux function $f(u)$, suppose that it has a (unique) sonic point at $u = u_s$. Then an alternative finite volume scheme has a form that replaces the flux function of (4.75) by

$$F_{j+1/2}^{n+1/2} = \tfrac{1}{2} \left[\left(1 + \operatorname{sgn} A_j^n\right) F_j^n + \left(\operatorname{sgn} A_{j+1}^n - \operatorname{sgn} A_j^n\right) f(u_s) \right.$$
$$\left. + \left(1 - \operatorname{sgn} A_{j+1}^n\right) F_{j+1}^n \right]. \tag{4.77}$$

This scheme, which uses the signs of the characteristic speeds $\{A_j^n = a(U_j^n)\}$ rather than those of the divided differences $\{A_{j+1/2}^n\}$, is due to Engquist and Osher[1] and has been very widely used and studied; it differs from the Roe scheme only when a sonic point occurs between U_j^n and U_{j+1}^n, it is also TVD (see Exercise 11) and it correctly resolves the transonic rarefaction wave.

However, these two schemes are only first order accurate and it is no easy matter to devise TVD schemes that are second order accurate. To consider why this is so let us consider an explicit TVD three-point scheme in the form (4.72) and satisfying the conditions (4.73). For the linear advection equation $u_t + au_x = 0$ we suppose that C and D are constants. Then it is easy to see, following the argument that led to the Lax–Wendroff method in (4.36), that second order accuracy leads directly to these coefficients, as in (4.76), and hence the violation of the TVD conditions except in very special cases. From another viewpoint, in our two successful TVD schemes we have constructed the fluxes from just the cell average values U_j^n in each cell, and we cannot expect

---

[1] Engquist, B. and Osher, O. (1981), One-sided difference approximations for nonlinear conservation laws, *Math. Comp.* **36**, 321–52.

to approximate the solution to second order accuracy with a piecewise constant approximation.

This observation points the way to resolving the situation: an intermediate stage, variously called *recovery* or *reconstruction*, is introduced to generate a higher order approximation $\tilde{U}^n(\cdot)$ to $u(\cdot, t_n)$ from the cell averages $\{U_j^n\}$. Probably the best known approach is that used by van Leer[1] to produce his MUSCL schemes (Monotone Upstream-centred Schemes for Conservation Laws). This uses discontinuous piecewise linear approximations to generate second order approximations. Another well-established procedure leads to the Piecewise Parabolic Method (PPM) scheme of Colella and Woodward,[2] which can be third order accurate. In all cases the recovery is designed to preserve the cell averages. So for the recovery procedure used in the MUSCL schemes, for each cell we need only calculate a slope to give a straight line through the cell average value at the centre of the cell, and this is done from the averages in neighbouring cells. The PPM, however, uses a continuous approximation based on cell-interface values derived from neighbouring cell averages: so a parabola is generated in each cell from two interface values and the cell average. In all such schemes only the interface fluxes are changed in the finite volume update procedure of (4.68). How this is done using the recovered approximation $\tilde{U}^n(\cdot)$ is beyond the scope of this book; we merely note that it is based on solving local evolution problems in the manner initiated by the seminal work of Godunov.[3] However, we must observe that to obtain a TVD approximation in this way it is necessary to place restrictions on the recovery process. A typical constraint is that it is *monotonicity preserving*; that is, if the $\{U_j^n\}$ are monotone increasing then so must be $\tilde{U}^n(\cdot)$.

4.8 The box scheme

To give some indication of the range of schemes that are used in practice we will describe two other very important schemes. The *box scheme* is a very compact implicit scheme often associated with the names of

[1] van Leer, B. (1979), Towards the ultimate conservative difference scheme V. A second order sequel to Godunov's method, *J. of Comput. Phys.* **32**, 101–36.

[2] Colella, P. and Woodward, P.R. (1984), The piecewise parabolic method (PPM) for gas-dynamical simulations, *J. of Comput. Phys.* **54**, 174–201.

[3] Godunov, S.K. (1959), A finite difference method for the numerical computation of discontinuous solutions of the equations of fluid dynamics. *Mat. Sb.* **47**, 271–306.

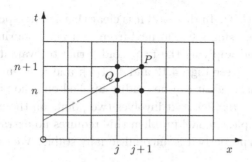

Fig. 4.12. The box scheme.

Thomée[1] and, in a different context, Keller:[2] for the simplest model problem $u_t + au_x = 0$ with constant a it takes the form

$$\frac{\delta_t \left(U_j^{n+1/2} + U_{j+1}^{n+1/2} \right)}{2\Delta t} + \frac{a\delta_x \left(U_{j+1/2}^n + U_{j+1/2}^{n+1} \right)}{2\Delta x} = 0. \qquad (4.78)$$

By introducing the averaging operator

$$\mu_x U_{j+1/2} := \tfrac{1}{2} \left(U_j + U_{j+1} \right), \qquad (4.79)$$

and similarly μ_t, we can write the scheme in the very compact form

$$(\mu_x \delta_t + \nu \mu_t \delta_x) U_{j+1/2}^{n+1/2} = 0, \qquad (4.80)$$

where $\nu = a\Delta t / \Delta x$ is the CFL number.

If we expand all the terms in Taylor series about the central point $(x_{j+1/2}, t_{n+1/2})$ as origin, it is easy to see that the symmetry of the averaged differences will give an expansion in even powers of Δx or Δt, so that the scheme is second order accurate. When the coefficient a is a function of x and t, it is sensible to replace it by $a_{j+1/2}^{n+1/2} :=$ $a(x_{j+1/2}, t_{n+1/2})$ in (4.78); this will leave the Taylor series unaltered, so that the truncation error remains second order.

When applied to the nonlinear problem in conservation form (4.46), it is written

$$\frac{\delta_t \left(U_j^{n+1/2} + U_{j+1}^{n+1/2} \right)}{2\Delta t} + \frac{\delta_x \left(F_{j+1/2}^n + F_{j+1/2}^{n+1} \right)}{2\Delta x} = 0 \qquad (4.81)$$

[1] Thomée, V. (1962), A stable difference scheme for the mixed boundary value problem for a hyperbolic first order system in two dimensions, *J. Soc. Indust. Appl. Math.* **10**, 229–45.

[2] Keller, H.B. (1971), A new finite difference scheme for parabolic problems, in B. Hubbard (ed.), *Numerical Solution of Partial Differential Equations II, SYNSPADE 1970*, Academic Press, pp. 327–50.

where $F_j^n := f(U_j^n)$. In this form it is clear that it corresponds to a finite volume scheme, using a box formed from four neighbouring mesh nodes in a square, and applying the trapezoidal rule to evaluate the integral along each edge – see Figs. 4.12 and 4.11(b); and in common with other finite volume schemes it can be readily applied on a nonuniform mesh.

The scheme is *implicit* as it involves two points on the new time level, but for the simplest model problem this requires no extra computation; and when used properly it is unconditionally stable. We can write (4.78) in the form

$$U_{j+1}^{n+1} = U_j^n + \left(1 + \nu_{j+1/2}^{n+1/2}\right)^{-1} \left(1 - \nu_{j+1/2}^{n+1/2}\right) \left(U_{j+1}^n - U_j^{n+1}\right) \quad (4.82)$$

where

$$\nu_{j+1/2}^{n+1/2} = a_{j+1/2}^{n+1/2} \frac{\Delta t}{\Delta x}.$$

When $a(x,t)$ is positive, so that the characteristic speed is positive, we must be given a boundary condition on the left of the region. This will define U_0^{n+1}, the first value of U on the new time level, and (4.82) will give directly the values of U in succession from left to right. If the speed is negative, and we are given a boundary condition on the right, a similar formula is used in the direction from right to left. When the equation is nonlinear and in conservation form the scheme is not quite so easy to use, as (4.81) now represents a nonlinear equation which must be solved for U_{j+1}^{n+1}.

One of the serious difficulties in using the scheme is the possibility of a *chequerboard mode* of the form $(-1)^{j+n}$ contaminating the solution. Because of the averaging in both space and time in equation (4.78) this is a spurious solution mode of this equation. It is only the boundary condition and the initial condition that control its presence; as we see in Fig. 4.13, it is much more evident with square pulse data than with smooth data.

For a system of equations, such as (4.5), the calculation becomes more elaborate, as the scheme is now truly implicit. We will have a set of simultaneous equations to solve on the new time level, and in general it will not be possible to solve them by a simple sweep in one direction, as it was for a single equation, since there will normally be some boundary conditions given at each end of the range of x. The situation is rather like that for implicit methods for the heat equation in Chapter 2, but not quite so straightforward. The matrix of the system will be typically block tridiagonal, with the block dimension equal to the order of

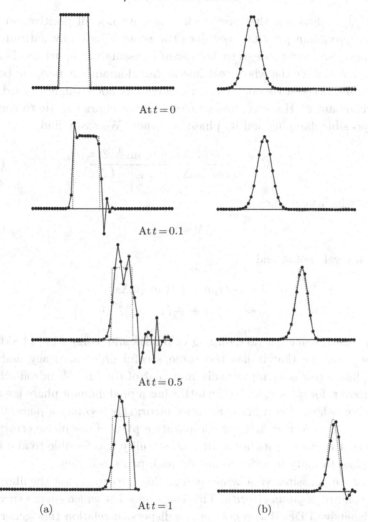

Fig. 4.13. Linear advection by the box scheme with $\Delta t = \Delta x = 0.02$ for (a) a square pulse and (b) Gaussian initial data.

the system, but the detailed structure will depend on the number of boundary conditions imposed at each end.

For the scalar problem the CFL condition is satisfied for any value of the ratio $\Delta t/\Delta x$ if we use the scheme in the correct direction, that is, in the form (4.82) when a is positive; as we see from Fig. 4.12, the characteristic always passes among the three points used in the construction of U_{j+1}^{n+1}. However, the three coefficients are not all positive, so there is no maximum principle; nor does the scheme have any natural TVD properties, being prone as we have seen to oscillatory solutions. Because we have neither the whole real line as our domain nor periodic boundary conditions, a rigorous Fourier analysis is not straightforward even for constant a. However, we can substitute a Fourier mode to consider its possible damping and its phase accuracy. We easily find

$$\lambda(k) = \frac{\cos \frac{1}{2}k\Delta x - i\nu \sin \frac{1}{2}k\Delta x}{\cos \frac{1}{2}k\Delta x + i\nu \sin \frac{1}{2}k\Delta x}, \qquad (4.83)$$

from which we deduce

$$|\lambda(k)| = 1 \qquad (4.84)$$

for any value of ν, and

$$\arg \lambda = -2\tan^{-1}\left(\nu \tan \tfrac{1}{2}k\Delta x\right)$$
$$\sim -\nu\xi\left[1 + \tfrac{1}{12}(1 - \nu^2)\xi^2 + \cdots\right]. \qquad (4.85)$$

Thus the scheme has no damping of modes and, comparing (4.85) with (4.40), we see that it has the same second order accuracy and that its phase error is asymptotically half that of the Lax–Wendroff scheme; moreover, for $|\nu| \leq 1$, while the latter has a predominant phase lag error, the box scheme has a phase advance error; this becomes a phase lag for $|\nu| > 1$. See Section 4.11 for comparative plots of the phase error; note too that because of its unconditional stability it is feasible to aim to set $|\nu|$ close to unity in order to obtain high phase accuracy.

For a nondissipative scheme such as this, we can exploit the illuminating concept of *group velocity*. This is the speed at which energy travels in a dispersive PDE: if $\omega = g(k)$ is the dispersion relation that generalises the relation (4.29) that holds for the linear advection equation, then the group velocity $C(k)$ is given by

$$C(k) := -\,\mathrm{d}\omega/\,\mathrm{d}k. \qquad (4.86)$$

So this reduces to $C(k) = a$ for linear advection. But for a difference scheme the group velocity is not so simple even in this case. If we write ω_h for $(\arg \lambda)/(\Delta t)$, then the relation given by (4.85) is the discrete dispersion relation

$$\tan \tfrac{1}{2}\omega_h\Delta t = -\nu \tan \tfrac{1}{2}k\Delta x.$$

Hence we can deduce a group velocity C_h for the scheme, corresponding to (4.86), given by

$$C_h = \nu\Delta x \sec^2 \tfrac{1}{2}k\Delta x/(\Delta t \sec^2 \tfrac{1}{2}\omega_h\Delta t)$$
$$= a \sec^2 \tfrac{1}{2}k\Delta x/(1 + \nu^2 \tan^2 \tfrac{1}{2}k\Delta x)$$
$$= \frac{a}{\cos^2 \tfrac{1}{2}\xi + \nu^2 \sin^2 \tfrac{1}{2}\xi}. \tag{4.87}$$

We could expand this directly in powers of $\xi = k\Delta x$ to obtain the error in the discrete group velocity for small ξ. However, it is more direct and more illuminating to use the expansion (4.85) that we already have for ω_h: differentiating this with respect to ξ we see immediately that the relative error is $\tfrac{1}{4}(1 - \nu^2)\xi^2$. This factor three increase, compared with the error in the relative phase speed, clearly comes from the second order accuracy of the scheme; and this simple relationship means that the group velocity adds little to our understanding of the behaviour of the scheme at low frequencies. However, there are general arguments that errors in a numerical scheme are composed of a range of frequencies so that the group velocity should be a better guide to a scheme's behaviour. In particular, if we look at the other end of the spectrum, at the oscillations for which $\xi = \pi$, we find that $C_h = a/\nu^2$. This again shows that these oscillations move ahead of the main wave when $0 < \nu < 1$ but the speed is more accurately predicted than is possible from the phase speed.

As regards stability, the box scheme is often regarded as unconditionally stable because of (4.84). But such a statement has to be made on the understanding that the equations are solved in the correct direction: clearly for $\nu < 0$ there would be an unbounded build-up of errors in applying the recursion (4.82); also such a form of calculation would then violate the CFL condition on domains of dependency. In practice, of course, this pitfall is readily avoided in most cases and the unconditional stability together with its compactness has commended the scheme to hydraulics engineers where it is widely used for river modelling.

We have already referred to the illustrative results that are displayed in Fig. 4.13(a,b). These are for the same linear advection problem (4.33) as before, with initial conditions, respectively, the square pulse and the Gaussian pulse. The discontinuities produce oscillations in the numerical solution that are considerably larger than those from the Lax–Wendroff scheme; but the accuracy is very similar for the smooth data. Notice the phase advance error in both cases since $|\nu| < 1$, and that the oscillations generated by the square pulse move at the speed predicted above by the group velocity. The oscillations are worse because of the lack of damping, but in practical problems where this might occur it is avoided by using the averaging $\theta U^{n+1} + (1 - \theta)U^n$ for the spatial difference in (4.78) with $\theta > \frac{1}{2}$.

From what has been said so far it may not be obvious why the scheme is so closely associated with river modelling, where it is generally called the Preissmann box scheme.[1] By integrating across a river channel, and introducing as dependent variables the cross-section area $A(x,t)$ and the total discharge $Q(x,t)$, one can derive the St Venant equations of river flow:

$$A_t + Q_x = 0$$
$$Q_t + (Q^2/A + \tfrac{1}{2}gA^2/b)_x = S, \qquad (4.88)$$

where x measures the distance along the river, S is a term that models the effect of the bed slope and frictional resistance, g is the gravitational constant and b the river breadth. The first equation just expresses mass conservation, and the second is the momentum equation. It is clear that conditions along the river will vary greatly with x. So one of the first attractions of the scheme is that the river can be divided into section, of different lengths, over each of which the river parameters are fairly constant. For example, the breadth may have considerable variations; also it may in general be related to A through the cross-section shape, though here we will assume a rectangular channel of breadth independent of A so that $A/b = h$, the river height. With this assumption the Jacobian matrix of the fluxes is easily calculated: in terms of h and the flow speed $u = Q/A$, it is just

$$A = \begin{pmatrix} 0 & 1 \\ gh - u^2 & 2u \end{pmatrix}, \qquad (4.89)$$

and is easily seen to have eigenvalues $u \pm c$ where $c = \sqrt{(gh)}$. In a normal slow-moving river the flow is sub-critical, i.e., $u < c$. Hence one characteristic speed is positive and one negative so that the pair of equations require one boundary condition on the left and one on the right; assuming the flow is from left to right, it is common for the discharge Q to be specified at the inflow on the left and the river height h at the outflow on the right. The system of equations to be solved is nonlinear and is solved by Newton iteration; as already mentioned, the resulting linear system is block tridiagonal with 2×2 blocks and can be solved by a matrix version of the Thomas algorithm.

---

[1] Preissmann, A. (1961), *Propagation des intumescences dans les canaux et rivières*. Paper presented at the First Congress of the French Association for Computation, Grenoble, France.

However, even with its compactness and the relatively straightforward procedure for solving the nonlinear system, it is still not clear why this implicit scheme should be used; and in view of the unremarkable phase accuracy indicated in (4.85) why is it that it is prized for the accuracy of its flood wave predictions? The answer lies in the structure of flood wave flows, and in particular the balance of the terms in the momentum equation. The forcing term on the right is usually written $S = g(S_0 - S_f)$, where $S_0 = S_0(x)$ represents the bed slope and $S_f = S_f(x, A, Q)$ the bed friction. Then the three terms in the equation which are multiplied by g are almost in balance. So this gives a relationship $Q = Q_g(x, A)$; indeed, in simpler models it is common to combine the mass balance equation with an experimentally observed relation of this kind to predict the flow. We can therefore deduce that a flood wave will move at a speed given by $\partial Q_g / \partial A$, which is typically not much larger than u and very much less than $u + c$. So the time step is chosen on the basis of this speed (with ν close to unity in (4.85) and hence giving accurate phase speeds) and the stability with respect to the characteristic speed $u + c$ is fully exploited. Also it is usual to take $\theta > \frac{1}{2}$ in the time averaging so as to damp the spurious mode oscillations, in a manner that will be studied in Section 5.8.

4.9 The leap-frog scheme

The second important scheme is called the *leap-frog scheme* because it uses two time intervals to get a central time difference and spreads its 'legs' to pick up the space difference at the intermediate time level; the values used are shown in Fig. 4.14. For (4.46) or (4.47) it has the form

$$\frac{U_j^{n+1} - U_j^{n-1}}{2\Delta t} + \frac{f(U_{j+1}^n) - f(U_{j-1}^n)}{2\Delta x} = 0, \qquad (4.90)$$

or

$$U_j^{n+1} = U_j^{n-1} - (a\Delta t/\Delta x)\left[U_{j+1}^n - U_{j-1}^n\right]. \qquad (4.91)$$

Thus it is an explicit scheme that needs a special technique to get it started. The initial condition will usually determine the values of U^0, but a special procedure is needed to give U^1. Then the leap-frog scheme can be used to give U^2, U^3,... in succession. The additional starting values U^1 can be obtained by any convenient one-step scheme, such as Lax–Wendroff.

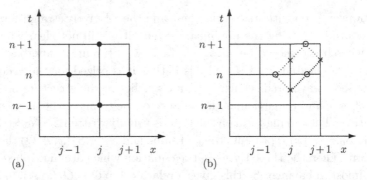

Fig. 4.14. The leap-frog scheme: (a) unstaggered; (b) staggered, × = V and o = W.

It is clear from Fig. 4.14(a) that the CFL condition requires that $|\nu| \le 1$, as for the Lax–Wendroff scheme. When $f = au$ with constant a the usual Fourier analysis leads to a quadratic for $\lambda(k)$:

$$\lambda^2 - 1 + 2i\nu\lambda \sin k\Delta x = 0 \tag{4.92}$$

with solutions

$$\lambda(k) = -i\nu \sin k\Delta x \pm \left[1 - \nu^2 \sin^2 k\Delta x\right]^{1/2}. \tag{4.93}$$

Since the product of these roots is -1, we must require both roots to have modulus 1 for the scheme to be stable. It is easy to verify that the roots are complex and equal in modulus for all k if and only if $|\nu| \le 1$: so for this scheme the Fourier analysis gives the same result as the CFL condition; and when the stability condition is satisfied there is no damping.

The result of the Fourier analysis leading to two values of $\lambda(k)$ is a serious problem for this scheme, as it means that it has a spurious solution mode. It arises from the fact that the scheme involves three time levels and so needs extra initial data, and it is this that determines the strength of this mode. Taking the positive root in (4.93) we obtain a mode that provides a good approximation to the differential equation, namely the 'true' mode λ_T given by

$$\arg \lambda_T = -\sin^{-1}\left(\nu \sin k\Delta x\right)$$
$$\sim -\nu\xi \left[1 - \tfrac{1}{6}(1 - \nu^2)\xi^2 + \cdots\right]. \tag{4.94}$$

Note that the phase error here has the same leading term as the Lax–Wendroff scheme – see (4.40). On the other hand, taking the negative root gives the spurious mode

$$\lambda_S \sim (-1)\left[1 + i\nu\xi - \tfrac{1}{2}\nu^2\xi^2 + \cdots\right],\qquad(4.95)$$

which gives a mode oscillating from time step to time step and travelling in the wrong direction. In practical applications, then, great care has to be taken not to stimulate this mode, or in some circumstances it may have to be filtered out.

The discrete dispersion relation for the scheme can be derived from (4.91), or from (4.92), as

$$\sin \omega_h \Delta t = -\nu \sin k\Delta x.$$

Differentiating this gives the group velocity, corresponding to (4.86),

$$C_h = \nu\Delta x \cos k\Delta x / \Delta t \cos \omega_h \Delta t$$
$$= \frac{a \cos k\Delta x}{(1 - \nu^2 \sin^2 k\Delta x)^{1/2}}.\qquad(4.96)$$

As with the box scheme, by expanding this expression or by differentiating (4.94), we deduce that the group velocity error of the true mode, i.e., for small $k\Delta x$, is three times the phase speed error given by (4.94). At the other extreme, the group velocity of the spurious or parasitic mode is obtained as $-a$ by setting $k\Delta x = \pi$ in this expression. Note that this is equivalent to considering the negative root of (4.93) in the limit $k\Delta x = 0$, since the two roots switch their roles when $k\Delta x$ passes through π as it traverses its allowed range $[0, 2\pi)$.

The results displayed in Fig. 4.15 illustrate the application of the leap-frog method for a square pulse and for Gaussian initial data; the first time step used the Lax–Wendroff scheme. The results clearly show the oscillating wave moving to the left. In some respects the results are similar to those for the box scheme; but the oscillations move at a speed independent of the mesh and cannot be damped, so in this case they have to be countered by some form of filtering.

The real advantage of the leap-frog method occurs when it is applied to a pair of first order equations such as those derived from the familiar second order wave equation

$$u_{tt} = a^2 u_{xx},\qquad(4.97)$$

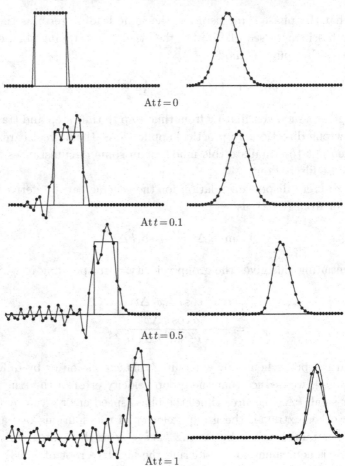

Fig. 4.15. Linear advection by the leap-frog scheme with $\Delta t = \Delta x = 0.02$ for (a) a square pulse and (b) Gaussian initial data.

where a is a constant: if we introduce variables $v = u_t$ and $w = -au_x$, it is clear that they satisfy the system

$$v_t + aw_x = 0,$$
$$w_t + av_x = 0. \tag{4.98}$$

Because of the pattern of differentials here, a staggered form of the leap-frog method can be used that is much more compact than (4.91): as indicated in Fig. 4.14(b) we have V and W at different points and a staggered scheme can be written

$$\frac{V_j^{n+1/2} - V_j^{n-1/2}}{\Delta t} + a\frac{W_{j+1/2}^n - W_{j-1/2}^n}{\Delta x} = 0, \qquad (4.99\text{a})$$

$$\frac{W_{j+1/2}^{n+1} - W_{j+1/2}^n}{\Delta t} + a\frac{V_{j+1}^{n+1/2} - V_j^{n+1/2}}{\Delta x} = 0, \qquad (4.99\text{b})$$

or

$$\delta_t V + \nu\delta_x W = 0, \qquad \delta_t W + \nu\delta_x V = 0, \qquad (4.100)$$

where we have taken advantage of the notation to omit the common superscripts and subscripts. With constant a, we can construct a Fourier mode by writing

$$(V^{n-1/2}, W^n) = \lambda^n e^{ikx}(\widehat{V}, \widehat{W}) \qquad (4.101)$$

where $\widehat{V}$ and $\widehat{W}$ are constants. These will satisfy the equations (4.99) if

$$\begin{pmatrix} \lambda - 1 & 2i\nu\sin\frac{1}{2}k\Delta x \\ 2i\lambda\nu\sin\frac{1}{2}k\Delta x & \lambda - 1 \end{pmatrix} \begin{pmatrix} \widehat{V} \\ \widehat{W} \end{pmatrix} = \begin{pmatrix} 0 \\ 0 \end{pmatrix}. \qquad (4.102)$$

This requires the matrix in (4.102) to be singular, so that

$$\lambda^2 - 2(1 - 2\nu^2\sin^2\tfrac{1}{2}k\Delta x)\lambda + 1 = 0 \qquad (4.103)$$

with solutions given by

$$\lambda_\pm(k) = 1 - 2\nu^2 s^2 \pm 2i\nu s[1 - \nu^2 s^2]^{1/2}, \qquad (4.104)$$

where $s = \sin\frac{1}{2}k\Delta x$. Again, for the scheme to be stable we need λ_+, λ_- to be a complex conjugate pair so that stability requires $|\nu| \leq 1$, in which case $|\lambda_\pm| = 1$. The phases are given by

$$\arg\lambda_\pm = \pm\sin^{-1}\left(2\nu s[1 - \nu^2 s^2]^{1/2}\right)$$
$$\sim \pm\nu\xi\left[1 - \tfrac{1}{24}(1 - \nu^2)\xi^2 + \cdots\right]. \qquad (4.105)$$

Note that the two roots of (4.103) are just the squares of the roots of (4.92) with Δx replaced by $\frac{1}{2}\Delta x$; hence the expansion in (4.105) corresponds to that in (4.94) with ξ replaced by $\frac{1}{2}\xi$. Both modes are now true modes which move to left and right at equal speeds, correctly approximating the behaviour of solutions to the wave equation. Note too that the accuracy is now better than that of the box scheme.

Substituting

$$V_j^{n+1/2} = (U_j^{n+1} - U_j^n)/\Delta t, \quad W_{j+1/2}^n = -a(U_{j+1}^n - U_j^n)/\Delta x$$

into the equations (4.99), (4.100) gives

$$(\delta_t^2 - \nu^2 \delta_x^2)U_j^n = 0, \qquad (4.106)$$

the simplest central difference representation of the second order wave equation (4.97) for U, together with a consistency relation. Note, too, that if we eliminate either V or W from the equations we find that both satisfy this second order equation. Some of the very attractive properties of this scheme will be derived, and put in a wider context, in the next section.

4.10 Hamiltonian systems and symplectic integration schemes

There are two important structural properties that lie behind the attractive features of the staggered leap-frog scheme applied to the wave equation: first, the wave equation (4.97) is the simplest example of a *Hamiltonian PDE*; and secondly, the staggered leap-frog scheme is one of the most common examples of a *symplectic integration scheme*. The importance of combining these two ideas has been most fully worked out over the last several years in the approximation of ordinary differential equation systems; but as they have recently been introduced into the area of PDEs we shall here outline what is involved, by means of the staggered leap-frog example. We will also show that the box scheme can share some of these properties. In doing so we shall mainly use the terminology and notation of Leimkuhler and Reich (2004).

Hamiltonian systems of ODEs have their origins in Hamilton's 1834 formulation of the equations of motion for a dynamical system, but have since been much generalised and their key properties widely studied. Let $\mathbf{q} \in \mathbb{R}^d$ and $\mathbf{p} \in \mathbb{R}^d$ be 'position' and 'momentum' variables, which together we will denote by $\mathbf{z}$, and $\mathcal{H}(\mathbf{q}, \mathbf{p}) \equiv \mathcal{H}(\mathbf{z}) : \mathbb{R}^d \times \mathbb{R}^d \to \mathbb{R}$, a smooth Hamiltonian function that defines the ODE system

$$\dot{\mathbf{z}} \equiv \begin{pmatrix} \dot{\mathbf{q}} \\ \dot{\mathbf{p}} \end{pmatrix} = \mathbf{J} \begin{pmatrix} \mathcal{H}_{\mathbf{q}} \\ \mathcal{H}_{\mathbf{p}} \end{pmatrix} \equiv \mathbf{J} \nabla_{\mathbf{z}} \mathcal{H}, \qquad (4.107)$$

where the *canonical structure matrix* $\mathbf{J}$ has the form

$$\mathbf{J} = \begin{pmatrix} 0 & I_d \\ -I_d & 0 \end{pmatrix} \quad \text{with} \quad \mathbf{J}^{-1} = \begin{pmatrix} 0 & -I_d \\ I_d & 0 \end{pmatrix},$$

in which I_d is the d-dimensional identity matrix. It is clear that $\mathcal{H}$ is constant along any trajectory; its value represents the energy of the system, which we shall sometimes denote by $E(\mathbf{z})$. Indeed, consider an arbitrary function $\mathcal{G} : \mathbb{R}^{2d} \to \mathbb{R}$, for which we will have, along any trajectory

$$\frac{d\mathcal{G}(\mathbf{z})}{dt} = (\nabla_{\mathbf{z}}\mathcal{G})^T \dot{\mathbf{z}} = (\nabla_{\mathbf{z}}\mathcal{G})^T \mathbf{J} \nabla_{\mathbf{z}}\mathcal{H} =: \{\mathcal{G}, \mathcal{H}\}. \qquad (4.108)$$

The expression $\{\mathcal{G}, \mathcal{H}\}$ is called the *Poisson bracket* of $\mathcal{G}$ and $\mathcal{H}$. It is clearly antisymmetric, and hence zero when $\mathcal{G} = \mathcal{H}$: and whenever it is identically zero the quantity $\mathcal{G}(\mathbf{q}, \mathbf{p})$ is constant along the trajectory. Then $\mathcal{G}$ is called a *constant of the motion*, with the energy being such a constant for any Hamiltonian system.

The best-known example of a Hamiltonian system is the simple plane pendulum, in which $d = 1$ and $\mathcal{H} = \frac{1}{2}p^2 - (g/L)\cos q$. The trajectories are given by $\dot{q} = p$, $\dot{p} = -(g/L)\sin q$, and from $\mathcal{H}(q, p) = \text{const.}$ along each, it is easy to deduce that in the (q, p)−phase plane they form the familiar closed curves around centres at $p = 0$, $q = 2m\pi$ separated by saddle points at $p = 0$, $q = (2m + 1)\pi$.

Of even greater significance than the existence of constants of the motion are the structural properties of the flow map formed from a set of trajectories of a Hamiltonian system: for example, in the scalar $d = 1$ case it is *area-preserving*; more generally it is said to be *symplectic*. To see what is involved in these ideas we need a few more definitions. A general mapping $\Psi : \mathbb{R}^{2d} \to \mathbb{R}^{2d}$ is said to be symplectic, with respect to the canonical structure matrix $\mathbf{J}$, if its Jacobian $\Psi_{\mathbf{z}}$ is such that

$$\Psi_{\mathbf{z}}^T \mathbf{J}^{-1} \Psi_{\mathbf{z}} = \mathbf{J}^{-1}. \qquad (4.109)$$

In the scalar case it is then easy to calculate that

$$\text{if} \quad \Psi_{\mathbf{z}} = \begin{pmatrix} a & b \\ c & d \end{pmatrix} \quad \text{then} \quad \Psi_{\mathbf{z}}^T \mathbf{J}^{-1} \Psi_{\mathbf{z}} = \begin{pmatrix} 0 & -ad + bc \\ ad - bc & 0 \end{pmatrix},$$

so that Ψ is simplectic iff $\det \Psi_{\mathbf{z}} \equiv ad - bc = 1$. Hence if this holds, and if $\mathbf{z} \in \Omega \subset \mathbb{R}^2$ is mapped into $\hat{\mathbf{z}} = \Psi(\mathbf{z}) \in \hat{\Omega} \subset \mathbb{R}^2$, we have

$$\int_{\hat{\Omega}} d\hat{\mathbf{z}} = \int_{\Omega} \det \Psi_{\mathbf{z}} \, d\mathbf{z} = \int_{\Omega} d\mathbf{z},$$

i.e., the mapping is area-preserving. So the symplectic property generalises the area preserving property to $d > 1$.

To apply this concept to the mapping produced by integrating a differential equation we define, in the language of differential geometry, the *differential one-form* of a function $f : \mathbb{R}^{2d} \to \mathbb{R}$, in the direction $\boldsymbol{\xi} \in \mathbb{R}^{2d}$,

$$\mathrm{d}f(\boldsymbol{\xi}) := \nabla_{\mathbf{z}} f \cdot \boldsymbol{\xi} \equiv \sum_{i=1}^{2d} \frac{\partial f}{\partial z_i} \xi_i. \tag{4.110}$$

Then for two such functions, f and g, we can define a *differential two-form*, called the *wedge product*, as

$$(\mathrm{d}f \wedge \mathrm{d}g)(\boldsymbol{\xi}, \boldsymbol{\eta}) := \mathrm{d}g(\boldsymbol{\xi}) \, \mathrm{d}f(\boldsymbol{\eta}) - \mathrm{d}f(\boldsymbol{\xi}) \, \mathrm{d}g(\boldsymbol{\eta}). \tag{4.111}$$

In particular, we can apply (4.110) to the components z_i of $\mathbf{z} \equiv (\mathbf{q}, \mathbf{p})$ to obtain $\mathrm{d}z_i(\boldsymbol{\xi}) = \xi_i$ and write these as a vector $\mathrm{d}\mathbf{z} \equiv (\mathrm{d}\mathbf{q}, \mathrm{d}\mathbf{p})^T = (\mathrm{d}z_1, \mathrm{d}z_2, \ldots, \mathrm{d}z_{2d})^T$. It is also easy to see that if we apply (4.110) to the components of the transformed variable $\hat{\mathbf{z}} = \Psi(\mathbf{z})$ we obtain

$$\mathrm{d}\hat{\mathbf{z}}(\boldsymbol{\xi}) = \Psi_{\mathbf{z}} \, \mathrm{d}\mathbf{z}(\boldsymbol{\xi}) \equiv \Psi_{\mathbf{z}} \boldsymbol{\xi}. \tag{4.112}$$

Furthermore, we can apply (4.111) to these components and then define the wedge product

$$\mathrm{d}\mathbf{q} \wedge \mathrm{d}\mathbf{p} := \sum_{i=1}^{d} \mathrm{d}q_i \wedge \mathrm{d}p_i. \tag{4.113}$$

It is the conservation of this quantity that turns out to be the key characterisation of Hamiltonian systems.

First of all, with a calculation as in the scalar case, we see that

$$\begin{aligned}
\boldsymbol{\xi}^T J^{-1} \boldsymbol{\eta} &= (\mathrm{d}\mathbf{q}^T(\boldsymbol{\xi}), \mathrm{d}\mathbf{p}^T(\boldsymbol{\xi})) J^{-1} (\mathrm{d}\mathbf{q}(\boldsymbol{\eta}), \mathrm{d}\mathbf{p}(\boldsymbol{\eta}))^T \\
&= \sum_{i=1}^{d} [\mathrm{d}p_i(\boldsymbol{\xi}) \, \mathrm{d}q_i(\boldsymbol{\eta}) - \mathrm{d}q_i(\boldsymbol{\xi}) \, \mathrm{d}p_i(\boldsymbol{\eta})] \\
&= \sum_{i=1}^{d} \mathrm{d}q_i \wedge \mathrm{d}p_i \equiv \mathrm{d}\mathbf{q} \wedge \mathrm{d}\mathbf{p}. \tag{4.114}
\end{aligned}$$

Then if we premultiply (4.109) by $\boldsymbol{\xi}^T$ and postmultiply by $\boldsymbol{\eta}$, and compare the result with the combination of (4.114) with (4.112), we deduce immediately that a mapping from $(\mathbf{q}, \mathbf{p})$ to $(\hat{\mathbf{q}}, \hat{\mathbf{p}})$ is symplectic iff

$$\mathrm{d}\hat{\mathbf{q}} \wedge \mathrm{d}\hat{\mathbf{p}} = \mathrm{d}\mathbf{q} \wedge \mathrm{d}\mathbf{p}. \tag{4.115}$$

The fundamental result that the flow map of a Hamiltonian system is symplectic can be derived directly from (4.109), but (4.115) is crucially important in characterising the behaviour of the flow.

Numerical methods for approximating ODE systems that retain these properties are called *symplectic integration schemes* or, more generally, *geometric integrators* – see Hairer, Lubich and Wanner (2002). The simplest of these share the staggered structure of the leap-frog scheme. For simplicity we start with the scalar $d = 1$ case, where we alternate between the pair of equations

$$q^{n+1} = q^n + \Delta t \, \mathcal{H}_p(q^n, p^{n+1/2})$$
$$p^{n+1/2} = p^{n-1/2} - \Delta t \, \mathcal{H}_q(q^n, p^{n+1/2}). \qquad (4.116)$$

If, as in the pendulum case, $\mathcal{H}_p$ depends only on p and $\mathcal{H}_q$ only on q this is an explicit method; more generally, it is implicit. In either case, if we take the differentials of these equations we obtain

$$dq^{n+1} = dq^n + \Delta t[\mathcal{H}_{pq} \, dq^n + \mathcal{H}_{pp} \, dp^{n+1/2}]$$
$$dp^{n+1/2} = dp^{n-1/2} - \Delta t[\mathcal{H}_{qq} \, dq^n + \mathcal{H}_{qp} \, dp^{n+1/2}], \qquad (4.117)$$

where we have omitted the arguments from the common Hamiltonian in (4.116). Now when we take the wedge product of these two equations, its antisymmetry implies that terms $dq^n \wedge \mathcal{H}_{qq} \, dq^n$ and $\mathcal{H}_{pp} \, dp^{n+1/2} \wedge dp^{n+1/2}$ are zero. So we take the wedge product of the first equation with $dp^{n+1/2}$ and substitute from the second equation in the dq^n term to get, after omitting these null terms,

$$dq^{n+1} \wedge dp^{n+1/2} = dq^n \wedge [\, dp^{n-1/2} - \Delta t \mathcal{H}_{qp} \, dp^{n+1/2}]$$
$$+ \Delta t \mathcal{H}_{pq} \, dq^n \wedge dp^{n+1/2}. \qquad (4.118)$$

The two terms in Δt cancel and we have the discrete symplectic property

$$dq^{n+1} \wedge dp^{n+1/2} = dq^n \wedge dp^{n-1/2}. \qquad (4.119)$$

If the whole procedure is repeated for a system with $d > 1$ the same result is obtained: this is because, from the definitions of (4.111) and (4.113), it is easy to see that for any matrix A we have

$$d\mathbf{a} \wedge (A \, d\mathbf{b}) = (A^T \, d\mathbf{a}) \wedge d\mathbf{b},$$

so that if A is symmetric and $\mathbf{a} = \mathbf{b}$ the antisymmetry of the wedge product again implies that the result is zero.

In the ODE literature this staggered leap-frog method is usually referred to as the Störmer-Verlet method; and the commonly used Asymmetrical

Euler methods differ from it only in their superscript labelling. Their effectiveness in the long time integration of Hamiltonian ODE systems is amply demonstrated in the references already cited.

The transfer of these ideas to PDEs is relatively recent; and there are several alternative approaches. One is to discretise in space so as to obtain a Hamiltonian system of ODEs to which the above ideas can be applied directly: there is increasing interest in mesh-free or particle methods to achieve this step, but as we have hitherto excluded particle methods we shall continue to do so here; alternatively, one may first make a discretisation in space and then apply the 'method of lines' to integrate in time, but we will not consider this here either. A more fundamental formulation is due to Bridges.[1] This leads to a *multi-symplectic* PDE which generalises the form of (4.107) to

$$\mathbf{K}\mathbf{z}_t + \mathbf{L}\mathbf{z}_x = \nabla_{\mathbf{z}} S(\mathbf{z}), \qquad (4.120)$$

where $\mathbf{K}$ and $\mathbf{L}$ are constant skew-symmetric matrices. Unfortunately, these matrices and linear combinations of them are often singular, and the formulation of a given system in this way not very obvious. We will therefore apply a more straightforward approach to a wave equation problem that generalises (4.97) and (4.98).

Suppose we have a Hamiltonian $\mathcal{H}(u,v)$ which is now an integral over the space variable(s) of a function of u, v and their spatial derivatives. Then to derive a Hamiltonian PDE we define a *variational derivative* of $\mathcal{H}$. For example, consider

$$\mathcal{H}(u,v) = \int E(x,t)\,\mathrm{d}x \equiv \int [f(u) + g(u_x) + \tfrac{1}{2}v^2]\,\mathrm{d}x, \qquad (4.121)$$

where we have not specified the interval on which u and v are defined and the equations are to hold; the integrand $E(x,t)$ is called the energy density. The variational derivative of a functional $\mathcal{G}(u)$ is defined by the relation

$$\int \delta_u \mathcal{G}(u)(\delta u)\,\mathrm{d}x = \lim_{\epsilon \to 0} \frac{\mathcal{G}(u + \epsilon \delta u) - \mathcal{G}(u)}{\epsilon};$$

[1] Bridges, T.J. (1997), Multi-symplectic structure and wave propagation, *Math. Proc. Camb. Philos. Soc.* **121**, 147–90.

and applying this to (4.121), with boundary conditions that ensure any boundary terms are zero, gives

$$\int \delta_u \mathcal{H}(u, v)(\delta u) \, dx = \lim_{\epsilon \to 0} \epsilon^{-1} \int [f(u + \epsilon \delta u) - f(u)$$

$$+ g((u + \epsilon \delta u)_x) - g(u_x)] \, dx$$

$$= \int [f'(u)\delta u + g'(u_x)(\delta u)_x] \, dx$$

$$= \int [f'(u) - \partial_x g'(u_x)]\delta u \, dx. \qquad (4.122)$$

Comparing the two sides we deduce that

$$\delta_u \mathcal{H}(u, v) = f'(u) - \partial_x g'(u_x). \qquad (4.123)$$

The the resulting Hamiltonian PDE is given as

$$\begin{pmatrix} u_t \\ v_t \end{pmatrix} = \begin{pmatrix} 0 & +1 \\ -1 & 0 \end{pmatrix} \begin{pmatrix} \delta_u \mathcal{H} \\ \delta_v \mathcal{H} \end{pmatrix}. \qquad (4.124)$$

That is,

$$u_t = v, \quad v_t = \partial_x g'(u_x) - f'(u). \qquad (4.125)$$

Moreover, from these equations we can deduce a local energy conservation law of the form $E_t + F_x = 0$: from differentiation of the terms in the energy density of (4.121) and substitution from (4.125) we get, after cancellation and collection of terms,

$$E_t = f'(u)v + g'(u_x)v_x + v[\partial_x g'(u_x) - f'(u)]$$

$$= [vg'(u_x)]_x =: -F_x. \qquad (4.126)$$

The quantity $F(x, t) = -vg'(u_x)$ is called the energy flux.

For example, let $f = 0$ and $g(u_x) = \frac{1}{2}(au_x)^2$ with constant a. Then (4.125) becomes

$$u_t = v, \quad v_t = a^2 u_{xx}, \qquad (4.127)$$

which is equivalent to the second order wave equation (4.97). If we set $w = -au_x$ we get the first order pair of equations (4.98) to which we applied the staggered leap-frog method in Section 4.9. Furthermore, since $vg'(u_x) = va^2 u_x = -avw$ the local energy conservation law becomes

$$[\tfrac{1}{2}v^2 + \tfrac{1}{2}w^2]_t + [avw]_x = 0, \qquad (4.128)$$

which we could deduce directly from (4.98). It is this local property that we shall now show is preserved in a discrete form by the staggered leap-frog scheme. It can be regarded as the simplest consequence of the symplectic character of the method, and corresponds to the energy being a constant of the motion in the ODE case. Consideration of wedge product relations of the form (4.119), which now have to be integrated or summed over the space variables, is beyond the scope of this account.

Using the compact notation of (4.100) and the averaging operator as defined in (4.79), we note that

$$(\mu_t V)(\delta_t V) \equiv \tfrac{1}{2}(V^+ + V^-)(V^+ - V^-) = \delta_t(\tfrac{1}{2}V^2).$$

Hence we obtain from this pair of equations

$$\delta_t[\tfrac{1}{2}(V^2 + W^2)] + \nu[(\mu_t V)(\delta_x W) + (\mu_t W)(\delta_x V)] = 0. \qquad (4.129)$$

The first term clearly represents a time-step difference of the discrete energy, with a division by Δt being provided from the factor ν multiplying the second term, and the second has terms proportional to a, V, W and $(\Delta x)^{-1}$; but it is not at all obviously a difference of energy fluxes. To see that it actually is so, we need to go back to the full form of the equations (4.99), restore the superscripts and subscripts and refer to the staggered mesh of Fig. 4.14(b). First we write out the terms in the energy difference ΔE in (4.129). It is given by

$$\Delta E = \tfrac{1}{2}\big[(V_j^{n+1/2})^2 + (W_{j+1/2}^{n+1})^2\big]$$
$$- \tfrac{1}{2}\big[(V_j^{n-1/2})^2 + (W_{j+1/2}^n)^2\big]. \qquad (4.130)$$

Then the corresponding flux difference ΔF in the equation can be written, after cancellation of the terms $V_j^{n+1/2} W_{j+1/2}^n$ and rearrangement of the others, as

$$a^{-1}\Delta F \equiv \tfrac{1}{2}\big(V_j^{n+1/2} + V_j^{n-1/2}\big)\big(W_{j+1/2}^n - W_{j-1/2}^n\big)$$
$$+ \tfrac{1}{2}\big(W_{j+1/2}^{n+1} + W_{j+1/2}^n\big)\big(V_{j+1}^{n+1/2} - V_j^{n+1/2}\big)$$
$$= \tfrac{1}{2}\big[V_j^{n-1/2}W_{j+1/2}^n + V_{j+1}^{n+1/2}W_{j+1/2}^n + V_{j+1}^{n+1/2}W_{j+1/2}^{n+1}\big]$$
$$- \tfrac{1}{2}\big[V_j^{n+1/2}W_{j+1/2}^{n+1} + V_j^{n+1/2}W_{j-1/2}^n + V_j^{n-1/2}W_{j-1/2}^n\big]. \qquad (4.131)$$

This looks fearfully complicated, but when one refers to Fig. 4.14(b) it has a very neat interpretation: if in each product one joins the mesh point corresponding to the V factor with that corresponding to the W factor, one sees that the terms in the first bracket give two sides of the 'diagonal' rectangle from $(j, n - \tfrac{1}{2})$ round to $(j + \tfrac{1}{2}, n + 1)$, while the other bracket gives the other two sides of the rectangle (shown by dotted lines in the figure).

Indeed, not only is (4.129) in the required local conservation form in terms of the difference (4.130) and (4.131), but it is best written as

$$\Delta E \Delta x + \Delta F \Delta t = 0: \qquad (4.132)$$

for then it represents an integral of the conservation law $E_t + F_x = 0$ over the diagonal rectangle, after application of the Gauss divergence theorem to give a line integral of $F \, dt - E \, dx$ around its perimeter. For instance, the integral of $\frac{1}{2}v^2$ from $(j+\frac{1}{2}, n+1)$ to $(j-\frac{1}{2}, n)$ and then on to $(j+\frac{1}{2}, n)$ gives the difference of $(V)^2$ terms in (4.130); while continuing the integral back to $(j + \frac{1}{2}, n + 1)$ gives no contribution because there is no net change in x. Similarly, the integral of avw from $(j + \frac{1}{2}, n)$ to $(j + \frac{1}{2}, n + 1)$ gives the $\frac{1}{2}V_{j+1}^{n+1/2}(W_{j+1/2}^n + W_{j+1/2}^{n+1})$ terms in (4.131), while the earlier section from $(j - \frac{1}{2}, n)$ to $(j + \frac{1}{2}, n)$ gives the combined first and last terms $\frac{1}{2}V_j^{n-1/2}(W_{j+1/2}^n - W_{j-1/2}^n)$.

The box scheme has also been shown to have multisymplectic properties.[1] So we will conclude this section by showing that it possesses a compact energy conservation law when applied to the wave equation system. In the compact notation used in (4.80) we can write the scheme as

$$\delta_t \mu_x V + \nu \delta_x \mu_t W = 0, \qquad \delta_t \mu_x W + \nu \delta_x \mu_t V = 0. \qquad (4.133)$$

Then, in the same way as in (4.129), we can deduce that

$$\delta_t \tfrac{1}{2}[(\mu_x V)^2 + (\mu_x W)^2)] + \nu[(\mu_t \mu_x V)(\delta_x \mu_t W) + (\mu_t \mu_x W)(\delta_x \mu_t V)] = 0. \qquad (4.134)$$

Now it is easy to check that

$$\begin{aligned}
(\mu_x A)(\delta_x B) + (\mu_x B)(\delta_x A) &\equiv \tfrac{1}{2}[(A^+ + A^-)(B^+ - B^-) \\
&\quad + (B^+ + B^-)(A^+ - A^-) \\
&= A^+ B^+ - A^- B^- \equiv \delta_x(AB).
\end{aligned} \qquad (4.135)$$

Hence we deduce that

$$\delta_t \tfrac{1}{2}[(\mu_x V)^2 + (\mu_x W)^2)] + \nu \delta_x[(\mu_t V)(\mu_t W)] = 0, \qquad (4.136)$$

which is the natural energy conservation law over the original mesh box.

4.11 Comparison of phase and amplitude errors

We return now to the use of Fourier analysis by which means we can compare all of the methods we have introduced in this chapter. As we

---

[1] Zhao, P.F. and Quin, M.Z. (2000), Multisymplectic geometry and multisymplectic Preissmann scheme for the KdV equation, *J. Phys. A* **33**, 3613–26.

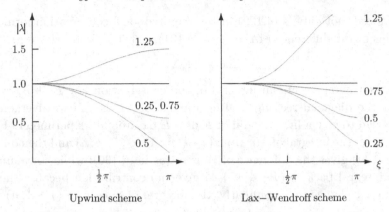

Fig. 4.16. Amplification factor $|\lambda|$ of Fourier modes, plotted against ξ, for $\nu = 0.25, 0.5, 0.75$ and 1.25. Note that for the upwind scheme the curves with $\nu = 0.25$ and $\nu = 0.75$ are identical.

have seen, the Fourier mode $u(x,t) = \mathrm{e}^{\mathrm{i}(kx+\omega t)}$ is an exact solution of the equation $u_t + au_x = 0$, for a a positive constant, provided that $\omega = -ak$. The amplitude of this mode is undamped, and in one time step its phase increases by $\omega \Delta t = -ak\Delta t$. A numerical scheme for this equation will have a solution of the form $\lambda^n \mathrm{e}^{\mathrm{i}kj\Delta x}$ where $\lambda(k)$ is a function of k, Δt and Δx. In one time step this mode is multiplied by the amplification factor λ, which is complex. The modulus of λ determines the stability of the scheme: if $|\lambda| > 1 + O(\Delta t)$ it will be unstable, and if $|\lambda| < 1$ the mode will be damped. The relative phase of the numerical solution to that of the exact solution is the ratio

$$\frac{\arg \lambda}{-ak\Delta t} = -\frac{\arg \lambda}{\nu \xi},$$

where $\xi = k\Delta x$ and $\nu = a\Delta t/\Delta x$.

Figure 4.16 shows graphs of $|\lambda|$ as a function of ξ for two of the four schemes which we have discussed, and Fig. 4.17 shows graphs of the relative phase for each of the four schemes. Both these quantities are close to unity when ξ is small, and their departure from it is a measure of the numerical errors in the schemes. We show the graphs over the range $0 \leq \xi \leq \pi$, though the phase error is important only for the interval $[0, \frac{1}{2}\pi]$ as the more rapidly oscillating modes are not properly represented on the mesh. The box scheme and the leap-frog scheme are undamped when $\nu < 1$, so these two graphs of $|\lambda|$ are omitted. For the

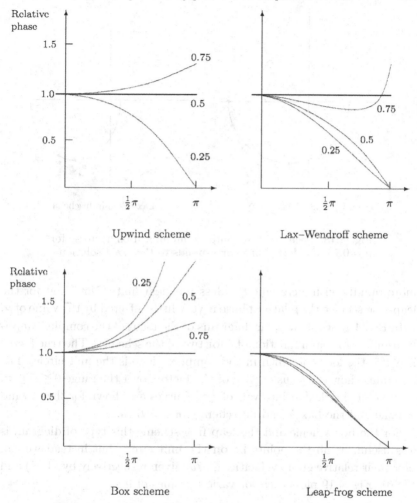

Fig. 4.17. Relative phase of Fourier modes, plotted against ξ, for $\nu = 0.25, 0.5$ and 0.75. The three curves shown for the leap-frog scheme are very close together; they increase with increasing ν.

upwind and Lax–Wendroff schemes we have included a curve with $\nu > 1$ to illustrate the behaviour in an unstable case.

Figure 4.17 shows clearly the phase advance of the box scheme for $\nu < 1$, and the phase lag of the Lax–Wendroff and leap-frog schemes. For small values of ξ the relative phases of the Lax–Wendroff and leap-frog schemes are very nearly the same; one can also see that the phase error of the box scheme is about half as large in magnitude, but it grows

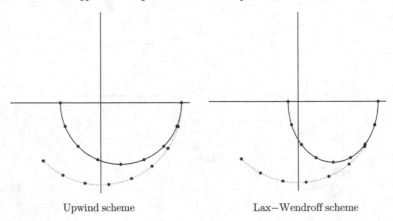

Upwind scheme   Lax−Wendroff scheme

Fig. 4.18. Locus of the complex amplification factors, for $\nu = 0.75$; the dotted line corresponds to the exact solution.

more rapidly with increasing ξ unless ν is near unity. Note that for the leap-frog scheme the relative phase is very little affected by the value of ν.

In Fig. 4.18 we show polar diagrams of the locus of the complex amplification factor λ as a function of ξ for two of the schemes. The true factor is $e^{-i\nu\xi}$, the locus of which in the complex plane is the unit circle. The diagrams show the locus of each of the factors over the range $0 \le \xi \le \pi$, with points marked at intervals of $\frac{1}{8}\pi$. Curves are shown for the upwind scheme and the Lax–Wendroff scheme for $\nu = 0.75$.

For the box scheme and the leap-frog scheme this type of diagram is less useful, as all the points lie on the unit circle; but instead we can plot their relative group velocities C_h/a, given respectively by (4.87) and (4.96). Fig. 4.19 plots each for various values of ν.

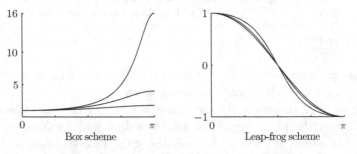

Box scheme   Leap-frog scheme

Fig. 4.19. Relative group velocity for the box and leap-frog schemes, for $\nu = 0.25, 0.5, 0.75$.

4.12 Boundary conditions and conservation properties

Both the Lax–Wendroff method and the unstaggered leap-frog method require an extra boundary condition over and above that associated with the differential equation. To calculate U_j^{n+1} requires values at $j-1$ and $j+1$ on the previous time level, which are only available when x_j is an internal point. Thus for the advection equation on the interval $(0,1)$ with $a > 0$ they need a condition on the right as well as that provided for the differential equation on the left. For systems of equations they need two boundary conditions for each equation rather than the one that would normally be given for the differential system. One of the advantages of the box scheme and the leap-frog method when it can be used in its staggered form is that they do not need these extra boundary conditions. Thus for the wave equation, which would have one boundary condition on the left and one on the right, the staggered leap-frog method is perfectly arranged if one of these specifies v and the other specifies w.

The question then arises as to how these extra boundary conditions should be derived when they are needed, and how they affect the properties of the scheme. Generally speaking, there are two ways of generating the boundary conditions and two ways of assessing their effect on the stability and accuracy of the difference scheme. The discrepancy in the need for boundary conditions on the part of the difference and differential systems is nearly always accounted for by the number of characteristics in the latter case that are pointing out of the domain. Thus the best way to derive the extra numerical boundary conditions is to approximate the corresponding characteristic equations. For example, with the advection equation one can always use the upwind scheme (4.20) to calculate the last point in a Lax–Wendroff or unstaggered leap-frog scheme.

For a system of equations, however, such a procedure would require finding the characteristic normal form of the system, which involves finding the eigenvalues and eigenvectors of the Jacobian matrix at the boundary point. An alternative, simpler, technique is just to set the first (or a higher order) difference of each unspecified variable equal to zero. A typical choice then on the right would be $U_J^n = U_{J-1}^n$.

The most precise means of analysing the effect of these extra boundary conditions is to consider their effect on the reflection and transmission of waves that move towards the boundary. We have seen how the unstaggered leap-frog method has spurious wave modes. Thus one wants to ensure that true modes travelling towards the boundary do not give rise to these spurious modes reflected back into the domain. Such analysis is very effective but rather complicated and a full discussion is beyond the scope of this book. We will, however, conclude this section with a description of a particular example of the technique.

An alternative approach is to try to choose boundary conditions so that conservation principles for the differential problem have analogues for the difference scheme. Thus if the differential equation (4.46) is solved on the unit interval with $u(0, t) = 0$ we obtain

$$\frac{\mathrm{d}}{\mathrm{d}t} \int_0^1 u(x, t)\, \mathrm{d}x = \int_0^1 u_t\, \mathrm{d}x = - \int_0^1 f_x\, \mathrm{d}x$$

$$= f(u(0, t)) - f(u(1, t)). \qquad (4.137)$$

For the two-step form of the Lax–Wendroff scheme (4.65), a corresponding relation would be

$$\Delta x \sum_1^{J-1} (U_j^{n+1} - U_j^n) = \Delta t \left[f(U_{1/2}^{n+1/2}) - f(U_{J-1/2}^{n+1/2}) \right] \qquad (4.138)$$

and the boundary conditions at both ends should be chosen so that (4.138) is a close analogue of (4.137). More will be said about these ideas in the next chapter.

The example of wave reflection and transmission at a boundary which we present is for the pair of equations $u_t + av_x = 0, v_t + au_x = 0$, with $a > 0$, approximated by the unstaggered leap-frog method, with a boundary on the left at $x = 0$. The wave equation system has two sets of solution modes, one corresponding to waves moving to the right, which we can write as

$$u_R = \mathrm{e}^{-\mathrm{i}k(x-at)}, \qquad v_R = \mathrm{e}^{-\mathrm{i}k(x-at)}; \qquad (4.139)$$

and the other corresponds to waves moving to the left, which we can write as

$$u_L = \mathrm{e}^{\mathrm{i}k(x+at)}, \qquad v_L = -\mathrm{e}^{\mathrm{i}k(x+at)}. \qquad (4.140)$$

If the boundary condition at $x = 0$ is $v = 0$, then we satisfy the differential equations and boundary condition with any solution of the form $u = A(u_R + u_L), v = A(v_R + v_L)$. Now we attempt to construct similar solutions to the leap-frog difference equations together with appropriate boundary conditions.

We cannot presume an exponential form for the mode and write instead

$$U_j^n = \lambda^n \mu^j \widehat{U}, \qquad V_j^n = \lambda^n \mu^j \widehat{V}, \qquad (4.141)$$

for some pair (λ, μ). Substituting these into the difference scheme,

$$U_j^{n+1} - U_j^{n-1} + \nu(V_{j+1}^n - V_{j-1}^n) = 0, \qquad (4.142a)$$
$$V_j^{n+1} - V_j^{n-1} + \nu(U_{j+1}^n - U_{j-1}^n) = 0, \qquad (4.142b)$$

where $\nu = a\Delta t/\Delta x$, gives the pair of algebraic equations

$$\begin{pmatrix} (\lambda^2 - 1)\mu & \nu\lambda(\mu^2 - 1) \\ \nu\lambda(\mu^2 - 1) & (\lambda^2 - 1)\mu \end{pmatrix} \begin{pmatrix} \widehat{U} \\ \widehat{V} \end{pmatrix} = \mathbf{0}. \qquad (4.143)$$

For these to have non-trivial solutions the pair (λ, μ) must be related by the determinantal equation

$$(\lambda^2 - 1)^2\mu^2 - \nu^2\lambda^2(\mu^2 - 1)^2 = 0. \qquad (4.144)$$

For fixed μ, both (4.143) and (4.144) are generalisations of what is obtained in a Fourier stability analysis, as for example in (4.102) and (4.103) for the staggered leap-frog scheme for the same pair of equations. However, in order to study the situation at the boundary, it is more useful to fix λ (and hence the variation with t) and consider the spatial modes given by the corresponding values for μ. From (4.144) we see that there are four solutions, arranged very symmetrically: rewriting (4.144) as

$$\mu^4 - 2\left[1 + \frac{(\lambda^2 - 1)^2}{2\nu^2\lambda^2}\right]\mu^2 + 1 = 0, \qquad (4.145)$$

we have a quadratic in μ^2 with roots whose product is unity; hence by writing $\gamma = (\lambda^2 - 1)/\nu\lambda$ we obtain

$$\mu = \pm\left[1 + \tfrac{1}{2}\gamma^2 - \gamma(1 + \tfrac{1}{4}\gamma^2)^{1/2}\right]^{\pm 1/2}. \qquad (4.146)$$

Suppose now we consider a low frequency mode that can be well approximated on the mesh and, by comparing with (4.139), we set $\lambda \approx e^{ika\Delta t} \approx 1 + ika\Delta t$ and hence get $\gamma \approx 2ik\Delta x$. For small values of γ, (4.146) gives

$$\mu \sim \pm(1 - \gamma)^{\pm 1/2} \quad \text{as } \gamma \to 0. \qquad (4.147)$$

Thus $(1 - \gamma)^{1/2} \approx e^{-ik\Delta x}$ and corresponds to a true right-moving wave, $(1 - \gamma)^{-1/2}$ corresponds to a true left-moving wave, and the two waves $-(1 - \gamma)^{\pm 1/2}$ are the two spurious modes that are associated with the leap-frog method. For each λ, we denote these roots by $\mu_{RT}, \mu_{LT}, \mu_{RS}, \mu_{LS}$ respectively, and have the relations

$$\mu_{RS} = -\mu_{RT}, \quad \mu_{LS} = -\mu_{LT} = -1/\mu_{RT}. \tag{4.148}$$

We can construct solutions with a given time variation from combinations of these four.

The eigenvectors corresponding to each μ are given from (4.143) as

$$\widehat{U} : \widehat{V} = (1 - \mu^2) : \mu\gamma, \tag{4.149}$$

and if the amplitudes of the two left-moving waves are given, the two required boundary conditions determine the amplitudes of the two right-moving waves. So suppose the true left-moving wave has unit amplitude and is uncontaminated by any spurious mode; then we have a solution to (4.142) of the form

$$\begin{pmatrix} U_j^n \\ V_j^n \end{pmatrix} = \lambda^n \left\{ \mu_{LT}^j \begin{pmatrix} \widehat{U}_{LT} \\ \widehat{V}_{LT} \end{pmatrix} + \alpha\mu_{RT}^j \begin{pmatrix} \widehat{U}_{RT} \\ \widehat{V}_{RT} \end{pmatrix} + \beta\mu_{RS}^j \begin{pmatrix} \widehat{U}_{RS} \\ \widehat{V}_{RS} \end{pmatrix} \right\}, \tag{4.150}$$

with α and β to be determined from the boundary conditions. The outcome that best matches the exact solution is to have $\alpha = 1, \beta = 0$ so that no spurious mode is generated by the wave reflection; this is achieved by a scheme introduced by Matsuno[1] in connection with numerical weather forecasting. As with the scheme (2.109) used for the Neumann condition applied to heat flow, and used to obtain heat conservation properties in Section 2.14, we set

$$U_0^n - U_1^n = 0, \quad V_0^n + V_1^n = 0. \tag{4.151}$$

The second condition approximates the given boundary condition $v = 0$, while the first is of the type which sets a difference to zero. Normalising to $\widehat{U}_{LT} = \widehat{U}_{RT} = \widehat{U}_{RS} = 1$ in (4.150), the first condition gives

$$(1 - \mu_{LT}) + \alpha(1 - \mu_{RT}) + \beta(1 - \mu_{RS}) = 0, \tag{4.152}$$

and substitution from (4.148) then gives

$$(1 - 1/\mu_{RT}) + \alpha(1 - \mu_{RT}) + \beta(1 + \mu_{RT}) = 0. \tag{4.153}$$

---

[1] Matsuno, T. (1966), False reflection of waves at the boundary due to the use of finite differences, *J. Meteorol. Soc. Japan* **44**(2), 145–57.

Similarly, the second boundary condition gives

$$(1 + 1/\mu_{RT})\widehat{V}_{LT} + \alpha(1 + \mu_{RT})\widehat{V}_{RT} + \beta(1 - \mu_{RT})\widehat{V}_{RS} = 0. \quad (4.154)$$

In this second equation, we substitute for $\widehat{V}$ from (4.149), making use of the identities

$$\frac{1/\mu}{1 - (1/\mu)^2} = -\frac{\mu}{1 - \mu^2}, \qquad \frac{-\mu}{1 - (-\mu)^2} = -\frac{\mu}{1 - \mu^2}, \qquad (4.155)$$

to get, after cancellation of the common factor $\gamma\mu_{RT}/(1 - \mu_{RT}^2)$,

$$-(1 + 1/\mu_{RT}) + \alpha(1 + \mu_{RT}) - \beta(1 - \mu_{RT}). \quad (4.156)$$

Finally, combining (4.153) with (4.156) gives

$$\alpha = 1/\mu_{RT}, \qquad \beta = 0, \quad (4.157)$$

which shows that no spurious right-moving wave is generated. Note that the true modes have the same amplitude at $j = \frac{1}{2}$, because $\mu_{LT}^{1/2} = \alpha\mu_{RT}^{1/2}$ follows from (4.153) and (4.148).

We have given this example in some detail partly because it is an analysis that can be carried out for other schemes (such as Lax–Wendroff, where the boundary conditions (4.151) are again very effective) and for other problems, but also because it is of the same type as that which is used for a general analysis of instabilities that can be generated by a poor choice of boundary conditions – see the Bibliographic notes of this chapter and Section 5.10 for references.

4.13 Extensions to more space dimensions

In some ways there is less difficulty in extending the methods of this chapter to two and three space dimensions than there was with the parabolic problems of Chapter 2 and Chapter 3. This is because it is less common and less necessary to use implicit methods for hyperbolic problems, since the stability condition $\Delta t = O(\Delta x)$ is less severe and to have Δt and Δx of similar magnitude is often required to maintain the accuracy of the schemes; also for important classes of problems, such as numerical weather prediction, the domain is periodic with no difficult curved boundaries to deal with. However, the theory of multi-dimensional hyperbolic equations is much less well developed and quite often the attractive properties that particular difference schemes have in one-dimension cannot be extended to two or more. Also, when it is advantageous to use implicit methods, the techniques for the fast solution of the resulting algebraic equations are much less easy to devise – see Chapter 7.

A typical system of equations can be written in a natural extension of (4.5) and (4.9) as

$$\frac{\partial u}{\partial t} + \frac{\partial f(u)}{\partial x} + \frac{\partial g(u)}{\partial y} = 0, \tag{4.158}$$

and

$$u_t + A(u)u_x + B(u)u_y = 0, \tag{4.159}$$

where the Jacobian matrices $A = \partial f/\partial u$, $B = \partial g/\partial u$ are derived from the flux functions f and g. The unstaggered leap-frog method which extends (4.90) can be written down immediately; and for the scalar case a Fourier analysis readily yields the stability condition

$$\left| \frac{a\Delta t}{\Delta x} \right| + \left| \frac{b\Delta t}{\Delta y} \right| \leq 1. \tag{4.160}$$

A staggered leap-frog scheme can be arranged in various ways, giving different stability conditions – see Exercise 10.

Similarly, the Lax–Wendroff method is easily extended and the way in which this is done will affect the stability conditions. For example, a widely used form of the two-step scheme can be written as follows, where we introduce the convenient averaging notation $\mu_x U_{i,j} := \frac{1}{2}(U_{i+1/2,j} + U_{i-1/2,j})$:

$$U^{n+1/2}_{i+1/2,j+1/2} = \left[\mu_x \mu_y U - \frac{1}{2} \left(\frac{\Delta t}{\Delta x} \mu_y \delta_x F + \frac{\Delta t}{\Delta y} \mu_x \delta_y G \right) \right]^n_{i+1/2,j+1/2}, \tag{4.161}$$

$$U^{n+1}_{i,j} = U^n_{i,j} - \left[\frac{\Delta t}{\Delta x} \mu_y \delta_x F + \frac{\Delta t}{\Delta y} \mu_x \delta_y G \right]^{n+1/2}_{i,j}. \tag{4.162}$$

In the scalar case this has the stability condition

$$\left(\frac{a\Delta t}{\Delta x} \right)^2 + \left(\frac{b\Delta t}{\Delta y} \right)^2 \leq 1, \tag{4.163}$$

which is the most natural multi-dimensional generalisation of a CFL condition for a wave with velocity vector (a, b).

As we have seen in Section 4.7 this scheme is a good example of a finite volume scheme; and the update (4.162) – which is of more general value than just for the Lax–Wendroff method – is readily generalised to a quadrilateral mesh such as those in Figs. 1.1 and 1.2. On such a mesh $\mathbf{U}_{i,j}^n$ is regarded as a cell average associated with the centroid of a cell and the fluxes in (4.162) need to be evaluated at the cell vertices. Suppose we denote these by a cyclic subscript α running in counter-clockwise order round the cell and apply the Gauss divergence theorem as in (4.66) to integrate the flux terms in (4.158) over the (i, j) cell. Using the trapezoidal rule for the edge integrals, the result is

$$|\Omega_{i,j}| \left(\mathbf{U}_{i,j}^{n+1} - \mathbf{U}_{i,j}^n \right) + \tfrac{1}{2} \Delta t \sum_\alpha \{ [\mathbf{F}_{\alpha+1} + \mathbf{F}_\alpha] (y_{\alpha+1} - y_\alpha)$$
$$- [\mathbf{G}_{\alpha+1} + \mathbf{G}_\alpha] (x_{\alpha+1} - x_\alpha) \}^{n+1/2} = 0.$$
$$(4.164)$$

Here the cell area $|\Omega_{i,j}|$ is given (by integrating $\mathrm{div}(x, 0)$ over the cell) as

$$|\Omega_{i,j}| = \tfrac{1}{2} \sum_\alpha (x_{\alpha+1} + x_\alpha)(y_{\alpha+1} - y_\alpha).$$

Note, too, that a similar formula could be derived for a triangular cell. So one could use a mesh composed of both quadrilaterals and triangles.

The more difficult step is that of calculating the vertex fluxes at the intermediate time level $n + \tfrac{1}{2}$. One could do so by generalising the Lax–Wendroff calculation given in (4.161) by integrating over a quadrilateral formed from the four cell centres around the vertex. But because of the oscillations associated with the use of Lax–Wendroff fluxes it is not clear that this would be very effective: one would rather extend to two dimensions the TVD flux constructs described in Section 4.7. However, for most cases of hyperbolic systems the matrices A and B do not commute. So they cannot be put into diagonal form simultaneously; and thus genuinely multi-dimensional forms of the upwind methods based on solving local Riemann problems are very difficult to construct.

Hence most multi-dimensional upwind schemes use fluxes calculated at interior points of the edges, rather than the vertices. Then one can use the fluxes in the normal direction, exploiting the property of hyperbolicity that every linear combination $\alpha A + \beta B$, where α and β are real, has real eigenvalues. The one-dimensional algorithms can therefore be applied directly, and the update step (4.164) has only to be modified by replacing the trapezoidal rule used on each edge by the mid-point rule or a Gauss rule.

Methods depend very much on the application and objectives being considered, and more detailed discussion is beyond the scope of the present book. Related difficulties arise at boundaries, where the correct form for boundary conditions cannot be deduced simply from the ideas of ingoing and outgoing characteristics as in one dimension.

Bibliographic notes and recommended reading

For an exposition of the basic theory of hyperbolic equations, characteristics, Riemann invariants and shocks see the classic texts by Courant and Hilbert (1962) and Courant and Friedrichs (1948), or rather more recent books by Carrier and Pearson (1976), Smoller (1983) and Evans (1998). For thorough expositions on nonlinear waves and waves in fluids see Whitham (1974) and Lighthill (1978).

Texts which combine a theoretical treatment of conservation laws and problems involving shocks with more details on their numerical modelling include those by LeVeque (1992, 2002) and by Kreiss and Lorenz (1989). The latter also contains an authoritative account of the well-posedness of initial-boundary-value problems for hyperbolic equations, and the closely related issue of numerical boundary conditions and their stability; an earlier text which pioneered the analysis exemplified in Section 4.12 is that by Godunov and Ryabenkii (1964).

Recent accounts of geometric and symplectic integration methods for Hamiltonian systems can be found in Hairer *et al.* (2002) and Leimkuhler and Reich (2004).

Exercises

4.1   Sketch the characteristics for the equation $u_t + au_x = 0$ for $0 \le x \le 1$ when $a \equiv a(x) = x - \frac{1}{2}$. Set up the upwind scheme on a uniform mesh $\{x_j = j\Delta x, j = 0, 1, \ldots, J\}$, noting that no boundary conditions are needed, and derive an error bound; consider both even and odd J. Sketch the development of the solution when $u(x, 0) = x(1-x)$ and obtain explicit error bounds by estimating the terms in the truncation error.

Repeat the exercise with $a(x) = \frac{1}{2} - x$, but with boundary conditions $u(0, t) = u(1, t) = 0$.

4.2   If q has an expansion in powers of p of the form

$$q \sim c_1 p + c_2 p^2 + c_3 p^3 + c_4 p^4 + \cdots ,$$

show that

$$\tan^{-1} q \sim c_1 p + c_2 p^2 + (c_3 - \tfrac{1}{3}c_1^3)p^3 + (c_4 - c_1^2 c_2)p^4 + \cdots$$

as in Lemma 4.1 in Section 4.4.

Use this result to derive the leading terms in the phase expansions of the following methods for approximating $u_t + au_x = 0$:

Upwind	$-\nu\xi + \tfrac{1}{6}\nu(1-\nu)(1-2\nu)\xi^3$;
Lax–Wendroff	$-\nu\xi + \tfrac{1}{6}\nu(1-\nu^2)\xi^3$;
Box	$-\nu\xi - \tfrac{1}{12}\nu(1-\nu^2)\xi^3$;
leap-frog	$-\nu\xi + \tfrac{1}{6}\nu(1-\nu^2)\xi^3$;

where $\nu = a\Delta t/\Delta x$ and $\xi = k\Delta x$.

4.3   Verify that the function $u(x,t)$ defined implicitly by the equation

$$u = f(x - ut)$$

is the solution of the problem

$$u_t + uu_x = 0, \qquad u(x,0) = f(x),$$

and that $u(x,t)$ has the constant value $f(x_0)$ on the straight line $x - x_0 = tf(x_0)$.

Show that the lines through the points $(x_0, 0)$ and $(x_0 + \epsilon, 0)$, where ϵ is small, meet at a point whose limit as $\xi \to 0$ is $(x_0 - f(x_0)/f'(x_0), -1/f'(x_0))$. Deduce that if $f'(x) \geq 0$ for all x the solution is single-valued for all positive t. More generally, show that if $f'(x)$ takes negative values, the solution $u(x,t)$ is single-valued for $0 \leq t \leq t_c$, where $t_c = -1/M$ and M is the largest negative value of $f'(x)$.

Show that for the function $f(x) = \exp[-10(4x-1)^2]$ the critical value is $t_c = \exp(\tfrac{1}{2})/8\sqrt{5}$, which is about 0.092. [Compare with Fig. 4.10.]

4.4   Determine the coefficients c_0, c_1, c_{-1} so that the scheme

$$U_j^{n+1} = c_{-1}U_{j-1}^n + c_0 U_j^n + c_1 U_{j+1}^n$$

for the solution of the equation $u_t + au_x = 0$ agrees with the Taylor series expansion of $u(x_j, t_{n+1})$ to as high an order as possible when a is a positive constant. Verify that the result is the Lax–Wendroff scheme.

In the same way determine the coefficients in the scheme

$$U_j^{n+1} = d_{-2}U_{j-2}^n + d_{-1}U_{j-1}^n + d_0U_j^n.$$

Verify that the coefficients d correspond to the coefficients c in the Lax–Wendroff scheme, but with ν replaced by $\nu-1$. Explain why this is so, by making the change of variable $\xi = x - \lambda t$ in the differential equation, where $\lambda = \Delta x/\Delta t$. Hence, or otherwise, find the stability conditions for the scheme.

4.5   For the scalar conservation law $u_t + f(u)_x = 0$, where $f(u)$ is a function of u only, the Lax–Wendroff scheme may be written

$$U_j^{n+1} = U_j^n - \frac{\Delta t}{\Delta x}P_j^n + \left(\frac{\Delta t}{\Delta x}\right)^2 Q_j^n,$$

where

$$P_j^n = \tfrac{1}{2}\left[f(U_{j+1}^n) - f(U_{j-1}^n)\right],$$

$$Q_j^n = \tfrac{1}{2}\left[A_{j+\frac{1}{2}}^n\left(f(U_{j+1}^n) - f(U_j^n)\right) - A_{j-\frac{1}{2}}^n\left(f(U_j^n) - f(U_{j-1}^n)\right)\right]$$

and

$$A_{j+\frac{1}{2}}^n = f'(\tfrac{1}{2}U_{j+1}^n + \tfrac{1}{2}U_j^n).$$

Expanding in Taylor series about the point (x_j, t_n) verify that the expansions of P and Q involve only odd and even powers of Δx respectively. Deduce that the leading terms in the truncation error of the scheme are

$$T_j^n = \tfrac{1}{6}(\Delta t)^2 u_{ttt} + (\Delta x)^2[\tfrac{1}{6}u_{xxx}f' + \tfrac{1}{2}u_x u_{xx}f'' + \tfrac{1}{6}u_x^3 f'''].$$

4.6   For the linear advection equation $u_t + au_x = 0$, where a is a positive constant, a generalised upwind scheme on a uniform mesh is defined by

$$U_j^{n+1} = (1 - \theta)U_k^n + \theta U_{k-1}^n$$

where $x_k - \theta\Delta x = x_j - a\Delta t$ and $0 \le \theta < 1$. Verify that the CFL condition requires no restriction on Δt, and that the von Neumann stability analysis also shows that stability is unrestricted. What is the truncation error of this scheme, and how does it behave as Δt increases?

4.7 Derive an explicit central difference scheme for the solution of

$$u_{xx} - (1 + 4x)^2 u_{tt} = 0$$

on the region $0 < x < 1, t > 0$, given

$$u(x, 0) = x^2, \quad u_t(x, 0) = 0, \quad u_x(0, t) = 0, \quad u(1, t) = 1.$$

Show how the boundary conditions are included in the numerical scheme. Find the characteristics of the differential equation, and use the CFL condition to derive a stability restriction.

4.8 The linearised one-dimensional forms of the isentropic compressible fluid flow equations are

$$\rho_t + q\rho_x + w_x = 0,$$
$$w_t + qw_x + a^2 \rho_x = 0,$$

where a and q are positive constants. Show that an explicit scheme which uses central differences for the x-derivatives is always unstable. By adding the extra terms arising from a Lax–Wendroff scheme, derive a conditionally stable scheme and find the stability condition.

4.9 Derive an 'angled-derivative' scheme for the advection equation $u_t + au_x = 0$, by combining the differences $U_j^{n+1} - U_j^n$ and $U_{j+1}^n - U_{j-1}^{n+1}$. Use Fourier analysis to study the accuracy and stability of the scheme. For $a > 0$, consider the boundary conditions that are needed to solve a problem on $0 \le x \le 1$ and the behaviour of the solution process at each time step.

4.10 For the wave equation in two dimensions,

$$\rho_t + u_x + v_y = 0, \quad u_t + c^2 \rho_x = 0, \quad v_t + c^2 \rho_y = 0,$$

set up a staggered leap-frog scheme with ρ at the points $x = rh$, $y = sh$ and (u, v) at the points $x = (r + \frac{1}{2})h$, $y = (s + \frac{1}{2})h$ on a square grid of size h, and find its stability condition.

Set up an alternative staggered leap-frog scheme in which ρ is at the same points but u is at the points $x = (r + \frac{1}{2})h$, $y = sh$ while v is at the points $x = rh$, $y = (s + \frac{1}{2})h$. Find its stability condition and compare with that of the former scheme, showing that one has a stability advantage over the other.

4.11   Show that the Engquist–Osher scheme of (4.77) is TVD on a
       uniform mesh when it is stable.

       Show that on a nonuniform mesh the Roe scheme of (4.74) is
       TVD if the time step is such that

$$-\Delta x_j \leq A_{j+1/2}^n \Delta t \leq \Delta x_{j+1},$$

       that is, a characteristic cannot travel from the cell edge $x_{j+1/2}$
       beyond the neighbouring cell edges in one time step.

5

Consistency, convergence and stability

5.1 Definition of the problems considered

In this chapter we shall gather together and formalise definitions that we have introduced in earlier chapters. This will enable us to state and prove the main part of the key *Lax Equivalence Theorem*. For simplicity we will not aim at full generality but our definitions and arguments will be consistent with those used in a more general treatment.

In the problems which we shall consider, we make the following assumptions:

- the region Ω is a fixed bounded open region in a space which may have one, two, three or more dimensions, with co-ordinates which may be Cartesian $(x, y, \ldots)$, cylindrical polar, spherical polar, etc.;

- the region Ω has boundary $\partial\Omega$;

- the required solution is a function u of the space variables, and of t, defined on $\Omega \times [0, t_F]$; this function may be vector-valued, so that our discussion can be applied to systems of differential equations, as well as to single equations;

- the operator $L(\cdot)$ involves the partial derivatives of u in the space variables; L does not involve t explicitly; for the most part we shall assume that L is a linear operator, but whenever possible we shall give definitions and state results which will generalise as easily as possible to nonlinear operators;

- the boundary conditions will prescribe the values of $g(u)$ on some or all of the boundary Ω, where $g(\cdot)$ is an operator which may involve spatial partial derivatives;

- the initial condition prescribes the value of u for $t = 0$ over the region Ω.

Thence we write the general form of the problems considered as

$$\frac{\partial u}{\partial t} = L(u) \quad \text{in } \Omega \times (0, t_F], \qquad (5.1a)$$

$$g(u) = g_0 \quad \text{on } \partial\Omega_1 \subset \partial\Omega, \qquad (5.1b)$$

$$u = u^0 \quad \text{on } \Omega \text{ when } t = 0. \qquad (5.1c)$$

We shall always assume that (5.1) defines a *well-posed problem*, in a sense which we shall define later; broadly speaking, it means that a solution always exists and depends continuously on the data.

5.2 The finite difference mesh and norms

Our finite difference approximation will be defined on a fixed mesh, with the time interval Δt constant both over the mesh and at successive time steps. The region Ω is covered by a mesh which for simplicity we shall normally assume has uniform spacing $\Delta x, \Delta y, \ldots$ in Cartesian co-ordinates, or $\Delta r, \Delta \theta, \ldots$ in polar co-ordinates. Individual values at mesh points will be denoted by U_j^n; in two or more space dimensions the subscript j will be used to indicate a multi-index, as a condensed notation for $U_{j,k}^n, U_{j,k,l}^n$, etc. We shall assume that a fixed, regular finite difference scheme is applied to a set of points where U_j^n is to be solved for and whose subscripts j lie in a set J_Ω, and it is only these points which will be incorporated in the norms. Usually this will be just the interior points of the mesh; and this means that where made necessary by curved boundaries, derivative boundary conditions etc., extrapolation to fictitious exterior points is used to extend the regular scheme to as many points as possible – see, e.g., Section 3.4 and the use of (3.35). There are other exceptional cases to consider too; when the regular finite difference operator is used at points on a symmetry boundary, as in Section 6.5 below, and also at points on boundaries where the boundary conditions are periodic, then these points are also included in J_Ω. The values of U at all such points on time level n will be denoted by U^n:

$$U^n := \{U_j^n, j \in J_\Omega\}. \qquad (5.2)$$

To simplify the notation we will consider schemes which involve only two time levels: for one-step methods this means that each U_j^n, if a vector, has the same dimension as u. However, as we have seen with the leap-frog method in Section 4.9, we can include multi-step methods by extending the dimension of U_j^n compared with u. For example, if a scheme involves three time levels, so that U^{n+1} is given in terms of

U^n and U^{n-1}, we can define a new vector $\tilde{U}^n$ with twice the dimension, whose elements are those of U^n and U^{n-1}.

To compare U with u we need to introduce norms which can be used on either, and in particular on their difference. Thus we first denote by u_j^n mesh values of the function $u(x,t)$ which will usually be the point values $u(x_j, t_n)$. We hope to show that the mesh values of U converge to these values of u. Then as for the mesh point values U_j^n above we define

$$u^n := \{u_j^n, j \in J_\Omega\}. \tag{5.3}$$

We shall consider just two norms. Firstly, the *maximum norm* is given by

$$\|U^n\|_\infty := \max\{|U_j^n|, j \in J_\Omega\}. \tag{5.4}$$

If we evaluate the maximum norm of u^n the result will approximate the usual supremum norm $\|u\|_\infty$ with u considered as a function of x at fixed time t_n, but will not in general be equal to it. The norms will only be equal if the maximum value of the function $|u(x, t_n)|$ is attained at one of the mesh points.

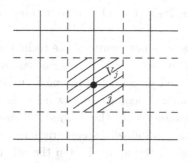

Fig. 5.1. Definition of control volume.

Secondly, we shall use a discrete l_2 norm which will approximate the integral L_2 norm. To do so, we introduce a 'control volume' with measure V_j associated with each interior mesh point: these will be non-overlapping elements whose union approximates Ω. Usually, as shown in Fig. 5.1, a mesh point x_j will lie at the centre of the control volume – see also Section 4.7 on finite volume methods; but this need not be the case so long as there is a one-to-one correspondence between mesh points and control volumes. In three-dimensional

Cartesian geometry, $V_j = \Delta x \Delta y \Delta z$; in three-dimensional cylindrical geometry, $V_j = r_j \Delta \theta \Delta r \Delta z$, and so on. Then, we define

$$\|U^n\|_2 := \left\{ \sum_{j \in J_\Omega} V_j |U_j^n|^2 \right\}^{1/2}. \tag{5.5}$$

For mesh points near the boundary the control volume may or may not be modified to lie wholly in Ω. In either case, the sum in (5.5) clearly approximates an integral so that $\|u^n\|_2$ approximates but does not in general equal the integral L_2 norm

$$\|u(\cdot, t_n)\|_2 := \left[\int_\Omega |u(x, t_n)|^2 \, dV \right]^{1/2} \tag{5.6}$$

at time t_n. However, if we define u_j^n as the root mean square value of $u(x, t_n)$ averaged over the jth control volume we clearly do have an exact match; we saw in Section 2.14 the value of making a similar interpretation when modelling heat conservation properties. For a single differential equation the notation $|U_j^n|$ is clear; if we are dealing with a system of differential equations, U_j^n is a vector and $|U_j^n|$ denotes a norm of this vector. The choice of which vector norm to use is immaterial to the subsequent analysis, but of course it must be used consistently throughout.

We should perhaps note here some of the techniques in common practical use that are not included in this general framework. Many of them have to do with adaptivity: choosing the next time step on the basis of a current error estimate; choosing a backward or forward difference in a solution dependent manner as in the upwind scheme of (4.20); or locally refining the mesh, e.g., to follow a steep gradient. Some of these would require major changes in the analysis. On the other hand, to cover the case of a refinement path composed of nonuniform meshes would not be very difficult.

5.3 Finite difference approximations

The general form of difference scheme we shall consider will be written

$$B_1 U^{n+1} = B_0 U^n + F^n. \tag{5.7}$$

As the notation implies, the difference operators B_0, B_1 are independent of n, corresponding to the assumption that $L(\cdot)$ does not depend explicitly on t; but, although based on fixed difference operators, they may depend on the point where they are applied. Thus at each point $j \in J_\Omega$, a linear difference operator B will be written in the form of a sum over near neighbours also in J_Ω:

$$(BU^n)_j = \sum_{k \in J_{\lambda l}} b_{j,k} U_k^n \quad \forall j \in J_\Omega; \tag{5.8}$$

while for a nonlinear operator, nonlinear combinations of U_k^n would be involved. The notation $b_{j,k}$ denotes the fact that the coefficients may depend on j as well as k, for two reasons. Firstly, it enables us to cover the case when $L(\cdot)$ has spatially variable coefficients, while for a constant coefficient problem, $b_{j,k}$ would usually just depend on the difference $k-j$. Secondly, although the pattern of neighbours involved in (5.8) will be the same for all points well away from the boundary, at the boundary we are assuming that the numerical boundary conditions have already been incorporated in (5.8) so that values of U at all points outside J_Ω have been eliminated. Thus the data term F^n in (5.7) includes not only data arising from inhomogeneous terms in the differential operator $L(u)$ but also inhomogeneous boundary data.

We shall always assume that B_1 is linear, of the form (5.8), so that it can be represented by a square matrix. To extend the theory to nonlinear problems it would be necessary for B_0 to be nonlinear but not necessarily B_1; but to cover schemes like the box scheme by such an extension would require B_1 to be nonlinear too.

We shall furthermore assume that B_1 is invertible, i.e. its representing matrix is non-singular. Hence we can write (5.7) as

$$U^{n+1} = B_1^{-1} [B_0 u^n + F^n]. \tag{5.9}$$

We shall also assume that (5.7) is so scaled that formally it represents the differential equation in the limit and hence $B_1 = O(1/\Delta t)$. Thus

$$B_1 u^{n+1} - [B_0 u^n + F^n] \to \frac{\partial u}{\partial t} - L(u) \tag{5.10}$$

as the mesh intervals $\Delta t, \Delta x, \ldots$ are refined in some manner which may depend on consistency conditions being satisfied. For example, in the θ-method (2.75) for the one-dimensional diffusion equation which we discussed in Chapter 2, away from the boundaries

$$B_1 = \frac{1}{\Delta t} - \theta \frac{\delta_x^2}{(\Delta x)^2}, \quad B_0 = \frac{1}{\Delta t} + (1 - \theta) \frac{\delta_x^2}{(\Delta x)^2}. \tag{5.11}$$

Moreover, we shall combine these two conditions and assume that the matrix B_1 is uniformly well-conditioned in the sense that there is a constant K such that, in whichever norm is being used to carry out the analysis,

$$\|B_1^{-1}\| \le K_1 \Delta t, \tag{5.12}$$

even though B_1^{-1} is represented by a matrix of ever-increasing dimension as the limit $\Delta t \to 0$ is approached.

For example, it is easy to deduce that in the case of (5.11) and the maximum norm, we have $K_1 = 1$: for the equation

$$B_1 U = F, \quad U = B_1^{-1} F \tag{5.13}$$

means, with $\mu = \Delta t / (\Delta x)^2$ and at a point away from the boundaries,

$$-\mu \theta U_{j-1} + (1 + 2\mu\theta) U_j - \mu \theta U_{j+1} = \Delta t F_j,$$

i.e.,

$$(1 + 2\mu\theta) U_j = \Delta t F_j + \mu\theta (U_{j-1} + U_{j+1}). \tag{5.14}$$

With Dirichlet boundary conditions, $B_1 U = F$ at $j = 1$ and $J - 1$ involves only two points of $J_\Omega \equiv \{j = 1, 2, \ldots, J - 1\}$, giving

$$(1 + 2\mu\theta) U_1 - \mu\theta U_2 = \Delta t F_1 \text{ and } -\mu\theta U_{J-2} + (1 + 2\mu\theta) U_{J-1} = \Delta t F_{J-1}$$

respectively. Thus for all values of $j \in J_\Omega$ we have

$$(1 + 2\mu\theta) |U_j| \le \Delta t \|F\|_\infty + 2\mu\theta \|U\|_\infty$$

and hence

$$(1 + 2\mu\theta) \|U\|_\infty \le \Delta t \|F\|_\infty + 2\mu\theta \|U\|_\infty \tag{5.15}$$

from which the result follows.

5.4 Consistency, order of accuracy and convergence

All limiting operations or asymptotic results refer (sometimes implicitly) to an underlying refinement path or set of refinement paths. That is, as in (2.47), a sequence of choices of the mesh parameters $\Delta t, \Delta x, \Delta y$, etc. is made such that they each tend to zero: and there may be inequality constraints between them. For brevity we shall characterise the whole of the spatial discretisation by a single parameter h: this may be just the largest of the mesh intervals $\Delta x, \Delta y, \ldots$, though this may need to be scaled by characteristic speeds in each of the co-ordinate directions; or h may be the diameter of the largest control volume around the mesh points. Then taking the limit along some designated refinement path we shall denote by '$\Delta t(h) \to 0$', or sometimes just $\Delta t \to 0$ or $h \to 0$: we shall always need Δt to tend to zero but stability or consistency may require that it does so at a rate determined by h, for example $\Delta t = O(h^2)$ being typical in parabolic problems and $\Delta t = O(h)$ in hyperbolic problems.

The *truncation error* is defined in terms of the exact solution u as

$$T^n := B_1 u^{n+1} - [B_0 u^n + F^n], \qquad (5.16)$$

and *consistency* of the difference scheme (5.7) with the problem (5.1a)–(5.1c) as

$$T_j^n \to 0 \quad \text{as } \Delta t(h) \to 0 \quad \forall j \in J_\Omega \qquad (5.17)$$

for all sufficiently smooth solutions u of (5.1a)–(5.1c). Note that this includes consistency of the boundary conditions through the elimination of the boundary values of U in the definition of B_0 and B_1.

If p and q are the largest integers for which

$$|T_j^n| \le C\left[(\Delta t)^p + h^q\right] \quad \text{as } \Delta t(h) \to 0 \quad \forall j \in J_\Omega \qquad (5.18)$$

for sufficiently smooth u, the scheme is said to have order of accuracy p in Δt and q in h: or pth order of accuracy in Δt, and qth order of accuracy in h.

Convergence on the other hand is defined in terms of all initial and other data for which (5.1a)–(5.1c) is well-posed, in a sense to be defined in the next section. Thus (5.7) is said to provide a *convergent approximation* to (5.1a)–(5.1c) in a norm $\| \cdot \|$ if

$$\|U^n - u^n\| \to 0 \quad \text{as } \Delta t(h) \to 0, n\Delta t \to t \in (0, t_F] \qquad (5.19)$$

for every u^0 for which (5.1a)–(5.1c) is well-posed in the norm: here we mean either of the norms (5.4) or (5.5). From a practical viewpoint the advantage of this approach is that the effect of round-off errors can be immediately allowed for: if on the other hand convergence were established only for sufficiently smooth data, round-off errors would have to be accounted for in a separate analysis.

5.5 Stability and the Lax Equivalence Theorem

None of the definitions (5.16)–(5.19) in the last section was limited to linear problems: they are quite general. In this section (and most of the rest of the chapter) however we are able to consider only linear problems. Suppose two solutions V^n and W^n of (5.7) or (5.9) have the same inhomogeneous terms F^n but start from different initial data V^0 and W^0: we say the scheme is *stable* in the norm $\| \cdot \|$ and for a given refinement path if there exists a constant K such that

$$\|V^n - W^n\| \le K\|V^0 - W^0\|, \quad n\Delta t \le t_F; \tag{5.20}$$

the constant K has to be independent of V^0, W^0 and of $\Delta t(h)$ on the refinement path, so giving a uniform bound. (In the nonlinear case, restrictions would normally have to be placed on the initial data considered.) The assumption that V^n and W^n have the same data F^n is a simplification that mainly stems from our decision to limit the consideration of boundary effects in the evolutionary problems treated in this and earlier chapters – this is why we have been able to set boundary conditions to zero when applying Fourier analysis to study stability. We shall study boundary effects much more in Chapter 6 on elliptic problems. Note too that we can also include the effect of interior data by application of Duhamel's principle, as was done in Section 2.11 of Chapter 2.

Since we are dealing with the linear case (5.20) can be written

$$\| \left(B_1^{-1} B_0 \right)^n \| \le K, \quad n\Delta t \le t_F. \tag{5.21}$$

Notice that for implicit schemes the establishment of (5.12) is an important part of establishing (5.21); consider, for example, the box scheme for linear advection, and the effect of having boundary conditions on one side or the other.

It is now appropriate to formalise our definition of well-posedness. We shall say that the problem (5.1) is *well-posed* in a given norm $\| \cdot \|$ if, for all sufficiently small h, we can show that (i) a solution exists for all data u^0 for which $\|u^0\|$ is bounded independently of h, and (ii) there exists a constant K' such that for any pair of solutions v and w,

$$\|v^n - w^n\| \le K'\|v^0 - w^0\|, \quad t_n \le t_F. \tag{5.22}$$

This differs from the usual definition in that we are using discrete norms; but we have chosen each of these so that it is equivalent to the corresponding function norm as $h \to 0$, if this exists for u, and we define u_j^n appropriately. An important feature of either definition is the following: for u to be a classical solution of (5.1a) it must be sufficiently smooth for the derivatives to exist; but suppose we have a sequence of data sets for which smooth solutions exist and these data sets converge to arbitrary initial data u^0 in the $\| \cdot \|$ norm, uniformly in h; then we can define a *generalised solution* with this data as the limit at any time t_n of the solutions with the smooth data, because of (5.22). Thus the existence of solutions in establishing well-posedness has only to be proved for a dense set of smooth data (with the definition of denseness again being uniform in h).

There is clearly a very close relationship between the definition of well-posedness for the differential problem and that of stability given by (5.20) for the discrete problem. This definition of stability, first formulated by Lax in 1953, enabled him to deduce the following key theorem:

Theorem 5.1 (Lax Equivalence Theorem) *For a consistent differ-ence approximation to a well-posed linear evolutionary problem, which is uniformly solvable in the sense of (5.12), the stability of the scheme is necessary and sufficient for convergence.*

Proof (of sufficiency). Subtracting (5.16) from (5.7) we have

$$B_1 \left(U^{n+1} - u^{n+1}\right) = B_0 \left(U^n - u^n\right) - T^n,$$

i.e.,

$$U^{n+1} - u^{n+1} = \left(B_1^{-1} B_0\right) \left(U^n - u^n\right) - B_1^{-1} T^n. \qquad (5.23)$$

Assuming that we set $U^0 = u^0$, it follows that

$$U^n - u^n = -[B_1^{-1} T^{n-1} + \left(B_1^{-1} B_0\right) B_1^{-1} T^{n-2} + \cdots$$
$$+ \left(B_1^{-1} B_0\right)^{n-1} B_1^{-1} T^0]. \qquad (5.24)$$

Now in applying the theorem, (5.12) and (5.21) are to hold in the same norm, for which we shall also deduce (5.19); we can combine these two to obtain

$$\| \left(B_1^{-1} B_0\right)^m B_1^{-1} \| \leq K K_1 \Delta t \qquad (5.25)$$

from which (5.24) gives

$$\|U^n - u^n\| \leq K K_1 \Delta t \sum_{m=0}^{n-1} \|T^m\|.$$

Thus convergence in the sense of (5.19) follows from the consistency of (5.17), if u is sufficiently smooth for the latter to hold. For less smooth solutions, convergence follows from the hypotheses of well-posedness and stability: general initial data can be approximated arbitrarily closely by data for smooth solutions and the growth of the discrepancy is bounded by the well-posedness of the differential problem and the stability (5.19) of the discrete problem.

The necessity of stability for convergence follows from the principle of uniform boundedness in functional analysis, working in the framework of a single Banach space for the continuous and discrete problems; this is where our simplified approach based on discrete norms has its disadvantages, and therefore a consideration of this principle is beyond the scope of this book – interested readers may find a proof tailored to this application inRichtmyer and Morton (1967), pp. 34–36, 46.   □

Thus for any scheme where consistency is readily established, we need only be concerned with establishing the conditions for stability; that is, we need only work with the discrete equations. As we have seen, consistency will usually hold for any sequence $\Delta t \to 0, h \to 0$; but there are a few cases where one has to be careful. For example, the Dufort–Frankel scheme for the one-dimensional heat equation,

$$\frac{U_j^{n+1} - U_j^{n-1}}{2\Delta t} = \frac{U_{j+1}^n - U_j^{n+1} - U_j^{n-1} + U_{j-1}^n}{(\Delta x)^2}, \qquad (5.26)$$

has the advantage of being explicit and yet is unconditionally stable, something contrary to our experience so far. However, one finds for the truncation error

$$T = (u_t - u_{xx}) + (\Delta t/\Delta x)^2 u_{tt} + O\big((\Delta t)^2 + (\Delta x)^2 + ((\Delta t)^2/\Delta x)^2\big). \quad (5.27)$$

Thus this is consistent with the heat equation only if $\Delta t = o(\Delta x)$ and is first order accurate only if $\Delta t = O((\Delta x)^2)$. As a result, it is the consistency condition rather than the stability condition that determines the refinement paths that can be used to obtain convergence.

This serves to emphasise the fact that, in the Lax Equivalence Theorem, there is implicit not only a choice of norm for defining stability and convergence but also a choice of refinement path.

5.6 Calculating stability conditions

As we are dealing with linear problems, if in (5.20) V^n and W^n are solutions of the difference equations (5.7), then the difference $V^n - W^n$ is a solution of the homogeneous difference equations with homogeneous boundary data. That is, establishing stability is equivalent to establishing the following:

$$B_1 U^{n+1} = B_0 U^n \text{ and } n\Delta t \leq t_F \quad \Rightarrow \quad \|U^n\| \leq K\|U^0\|, \qquad (5.28)$$

which is what is meant by (5.21). The constant K will generally depend on the time interval t_F and allows for the sort of exponential growth that might occur with $u_t = u_x + u$, for example. For simple problems one will often find: either $K = 1$, there is no growth and the scheme is stable; or $U^n \sim \lambda^n U^0$, with $|\lambda| > 1$ even as $\Delta t \to 0$ for some mode, so the scheme is unstable.

Thus when we established a maximum principle in Section 2.6 and elsewhere we were also establishing stability in the maximum norm: strictly speaking, we had also to establish a minimum principle so as to be able to say not only

$$U_j^{n+1} \leq \max_k U_k^n \leq \|U^n\|_\infty \qquad (5.29)$$

but also

$$U_j^{n+1} \geq \min_k U_k^n \geq -\|U^n\|_\infty \qquad (5.30)$$

and could then deduce

$$\|U^{n+1}\|_\infty \leq \|U^n\|_\infty. \qquad (5.31)$$

For parabolic problems proving stability by this means is very natural because a maximum (or minimum) principle is a very attractive and appropriate attribute for a difference scheme to possess over and above stability. Also, as we have seen in examples, quite general linear problems with variable coefficients and mixed boundary conditions can be dealt with: in each case we were able to deduce simple algebraic conditions on Δt for which a scheme could be shown to have a maximum principle and hence to be stable in the maximum norm. In many cases Fourier analysis could then show that, for the corresponding problem with constant coefficients and periodic boundary conditions, failing to satisfy these conditions leads to instability which, as we shall see below, is then in the l_2 norm. However, in other cases as with the θ-method, there was a gap between the two conditions given by the two methods. In an influential paper in 1952,[1] Fritz John showed that, for a wider class of parabolic problems and corresponding difference approximations, schemes that satisfy a von Neumann condition obtained from a locally applied Fourier analysis are also stable in the maximum norm. Thus, some schemes which do not satisfy a maximum principle are actually stable in the maximum norm; although see below in Section 5.7 for more comment on these cases.

[1] John, F. (1952) On the integration of parabolic equations by difference methods, *Comm. Pure Appl. Math.* **5**, 155.

Furthermore, a maximum principle is seldom available or even appropriate for hyperbolic problems. As we have noted, the first order scheme (4.20) satisfies a maximum principle whenever $0 \leq \nu \leq 1$ so that it is then stable in the maximum norm: but we can show that this can never be true of a second order scheme. For example, consider the Lax–Wendroff method written in the form (4.36). If it were to satisfy a maximum principle, then for any set of non-positive values for U^n one should never have $U^{n+1} > 0$: yet if $0 < \nu < 1$, setting $U_{j-1}^n = U_j^n = 0$ and $U_{j+1}^n = -1$ gives a positive value for U_j^{n+1}. This does not of course demonstrate that the scheme is actually unstable in the maximum norm, merely that we cannot prove such stability by this means.

For this reason, and also because hyperbolic differential equations are much more commonly well-posed in the L_2 norm than in the supremum norm, for hyperbolic problems we have to adopt the more modest target of proving stability in the l_2 norm (5.5). This gives weaker results because we have, recalling that V_j is the measure of the jth control volume,

$$\left[\min_{j \in J_\Omega} V_j\right]^{1/2} \|U\|_\infty \leq \|U\|_2 \leq \left[\sum_{j \in J_\Omega} V_j\right]^{1/2} \|U\|_\infty; \qquad (5.32)$$

in the bounded region we are working with, the coefficient on the right is a finite constant while that on the left tends to zero as the mesh is refined. It is clear that we would prefer to derive a maximum norm error bound from a stability analysis but, if we have only l_2 stability and so obtain a bound for the l_2 norm of the error $\|E^n\|_2$, then (5.32) gives a poor result for $\|E^n\|_\infty$.

However, it is the l_2 norm which is appropriate for Fourier analysis because of Parseval's relation. Suppose we can assume periodicity on a normalised region $[-\pi, \pi]^d$ which is covered by a uniform (Cartesian) mesh of size $\Delta x_1 = \Delta x_2 = \ldots = \Delta x_d = \pi/J$. Then the Fourier modes that can be distinguished on the mesh correspond to wave numbers, which we denote by the vector $\mathbf{k}$, having components given by

$$k = 0, \pm 1, \pm 2, \ldots, \pm J, \qquad (5.33)$$

where the last two with $k\Delta x = \pm \pi$ are actually indistinguishable. Hence we can expand any periodic function on the mesh as

$$U(\mathbf{x}_j) = \frac{1}{(2\pi)^{d/2}} \sum_{(\mathbf{k})}{}' \widehat{U}(\mathbf{k}) e^{i\mathbf{k} \cdot \mathbf{x}_j} \qquad (5.34)$$

where the prime on the summation sign means that any term with $k_s = \pm J$ has its weight halved, and we have also used a vector notation $\mathbf{x}_j$ for mesh points. This discrete Fourier expansion has an inverse which is the discrete Fourier transform

$$\widehat{U}(\mathbf{k}) = \frac{1}{(2\pi)^{d/2}} \sum{}'_{(j)} (\Delta x)^d U(\mathbf{x}_j) e^{-i\mathbf{k}\cdot\mathbf{x}_j}, \qquad (5.35)$$

where each component of j runs from $-J$ to J with the mesh points on the periodic boundaries again having their weights halved so that all the weights are equal to the V_j introduced in (5.5).

Lemma 5.1 *The Fourier modes* $(2\pi)^{-d/2}e^{i\mathbf{k}\cdot\mathbf{x}_j}$ *with components given by (5.33) are orthonormal with respect to the l_2 inner product used in (5.35), namely*

$$\langle U, W \rangle_2 := (\Delta x)^d \sum{}'_{(j)} U_j \overline{W}_j. \qquad (5.36)$$

Proof It is sufficient to consider $d = 1$. We first establish the fundamental trigonometric identity

$$\tfrac{1}{2}e^{-iJ\theta} + e^{-i(J-1)\theta} + \cdots + e^{i(J-1)\theta} + \tfrac{1}{2}e^{iJ\theta} = \sin J\theta \cot \tfrac{1}{2}\theta. \quad (5.37)$$

From the summation

$$1 + e^{i\theta} + e^{i2\theta} + \cdots + e^{i(J-1)\theta} = (e^{iJ\theta} - 1)/(e^{i\theta} - 1),$$

we obtain by adding $\tfrac{1}{2}(e^{iJ\theta} - 1)$

$$\tfrac{1}{2} + e^{i\theta} + e^{i2\theta} + \cdots + e^{i(J-1)\theta} + \tfrac{1}{2}e^{iJ\theta} = \tfrac{1}{2}(e^{iJ\theta} - 1)\frac{(e^{i\theta} + 1)}{(e^{i\theta} - 1)} \quad (5.38)$$

$$= \tfrac{1}{2i}(e^{iJ\theta} - 1)\cot \tfrac{1}{2}\theta. \quad (5.39)$$

Combining this with a similar sum for $-\theta$ gives (5.37). Now apply this with $\theta = (k_1 - k_2)\Delta x$, so that $J\theta = (k_1 - k_2)\pi$. We obtain

$$\sum{}'_{(j)} e^{ik_1 x_j} e^{-ik_2 x_j} = \sin(k_1 - k_2)\pi \cot \tfrac{1}{2}(k_1 - k_2)\Delta x, \quad k_1 \neq k_2,$$

so that

$$\sum{}'_{(j)} e^{ik_1 x_j} e^{-ik_2 x_j} = (2\pi/\Delta x)\delta_{k_1,k_2}.$$

$\square$

Hence we have, with V_j the control volume measure,

$$\|U\|_2^2 = \sum_{j \in J_\Omega} V_j |U_j|^2 \equiv \sideset{}{'}\sum_{(j)} (\Delta x)^d |U(\mathbf{x}_j)|^2$$

$$= \left(\frac{\Delta x}{2\pi}\right)^d \sideset{}{'}\sum_{(\mathbf{k})} |\widehat{U}(\mathbf{k})|^2 \left(\frac{2\pi}{\Delta x}\right)^d, \quad (5.40)$$

i.e.,

$$\|\widehat{U}\|_2^2 := \sideset{}{'}\sum_{(\mathbf{k})} |\widehat{U}(\mathbf{k})|^2 = \|U\|_2^2, \quad (5.41)$$

which is the appropriate form of Parseval's relation.

For a rectangular region of general dimensions a simple scaling will reduce the situation to the above case. However, note that not only is Δx then changed but we will also generally have $\Delta k \neq 1$ and that such a coefficient will be needed in the definition of $\|\widehat{U}\|_2$ for (5.41) to hold. It is also worth noting that when for example we have a problem on $[0, 1]$ with $u(0) = u(1) = 0$ we extend this to a periodic problem on $[-1, 1]$ by imposing antisymmetry at $x = 0$ and using a sine series. This is why we have taken $[-\pi, \pi]$ as our standard case above.

To establish (5.28) then, for a constant coefficient problem with periodic boundary conditions, we expand arbitrary initial data in the form (5.34) and, from the discrete Fourier transform of (5.28), obtain the same form at successive time levels with the coefficients given by

$$\widehat{B}_1(\mathbf{k})\widehat{U}^{n+1}(\mathbf{k}) = \widehat{B}_0(\mathbf{k})\widehat{U}^n(\mathbf{k}), \quad (5.42)$$

where, if the U^n are p-dimensional vectors, $\widehat{B}_0$ and $\widehat{B}_1$ are $p \times p$ matrices. The matrix

$$G(\mathbf{k}) = \widehat{B}_1^{-1}(\mathbf{k})\widehat{B}_0(\mathbf{k}) \quad (5.43)$$

is called the *amplification matrix* as it describes the amplification of each mode by the difference scheme. Because we have assumed that $\widehat{B}_0$ and $\widehat{B}_1$ are independent of t we can write

$$\widehat{U}^n = [G(\mathbf{k})]^n \widehat{U}^0 \quad (5.44)$$

and using (5.41) have

$$\sup_{U^0} \frac{\|U^n\|_2}{\|U^0\|_2} = \sup_{\widehat{U}^0} \frac{[\sum'_{(\mathbf{k})} |\widehat{U}^n(\mathbf{k})|^2]^{1/2}}{[\sum'_{(\mathbf{k})} |\widehat{U}^0(\mathbf{k})|^2]^{1/2}}$$

$$= \sup_{\mathbf{k}} \sup_{\widehat{U}^0(\mathbf{k})} \frac{|\widehat{U}^n(\mathbf{k})|}{|\widehat{U}^0(\mathbf{k})|} = \sup_{\mathbf{k}} |[G(\mathbf{k})]^n|. \quad (5.45)$$

Thus stability in the l_2 norm is equivalent to showing that

$$|[G(\mathbf{k})]^n| \leq K \quad \forall \mathbf{k}, \quad n\Delta t \leq t_F. \tag{5.46}$$

Here $|G^n|$ means the $p \times p$ matrix norm subordinate to the vector norm used for U_j^n and $\widehat{U}(\mathbf{k})$.

Then clearly we have the following result.

Theorem 5.2 (von Neumann Condition) *A necessary condition for stability is that there exist a constant K' such that*

$$|\lambda(\mathbf{k})| \leq 1 + K'\Delta t \quad \forall \mathbf{k}, \quad n\Delta t \leq t_F, \tag{5.47}$$

for every eigenvalue $\lambda(\mathbf{k})$ of the amplification matrix $G(\mathbf{k})$.

Proof By taking any eigenvector of $G(\mathbf{k})$ as $\widehat{U}(\mathbf{k})$ it is obviously necessary that there be a constant K such that $|\lambda^n| \leq K$: then by taking $n\Delta t = t_F$ we have

$$|\lambda| \leq K^{\Delta t/t_F} \leq 1 + (K-1)\Delta t/t_F \text{ for } \Delta t \leq t_F,$$

the last inequality following from the fact that K^s is a convex function of s. □

The von Neumann condition is a sufficient condition for the l_2 stability of any scheme to which the Fourier analysis can be applied, if the scheme is a one-step scheme for a scalar equation so that G is scalar. It is also sufficient if G is a normal matrix so that its subordinate norms can be bounded by its spectral radius. Other more sophisticated sufficient conditions are given in Richtmyer and Morton (1967), Chapter 4, as well as necessary and sufficient conditions derived by Kreiss[1] and Buchanan[2] in 1962 and 1963 respectively. The key result is known as the *Kreiss*

---

[1] Kreiss, H.O. (1962), Über die Stabilitätsdefinition für Differenzengleichungen die partielle Differentialgleichungen approximieren, *Nordisk Tidskr. Informations-Behandlung* **2**, 153–181.

[2] Buchanan, M.L. (1963), A necessary and sufficient condition for stability of difference schemes for second order initial value problems, *J. Soc. Indust. Appl. Math.* **11**, 474–501; and Buchanan, M.L. (1963), A necessary and sufficient condition for stability of difference schemes for initial value problems, *J. Soc. Indust. Appl. Math.* **11**, 919–35.

Matrix Theorem; its consideration is beyond the scope of this book, but see Section 5.9 below. However, it is worth noting here that, although, as we have seen, when the von Neumann condition is violated the growth of U^n can be exponential in n, when it is satisfied and G is uniformly bounded the growth can be at most polynomial in n: indeed at worst it is $O(n^{p-1})$ where G is a $p \times p$ matrix.

5.7 Practical (strict or strong) stability

Clearly the von Neumann condition is very important both practically and theoretically. Even for variable coefficient problems it can be applied locally (with local values of the coefficients) and because instability is a local phenomenon, due to the high frequency modes being the most unstable, it gives necessary stability conditions which can often be shown to be sufficient. However, for some problems the presence of the arbitrary constant in (5.47) is too generous for practical purposes, though being adequate for eventual convergence.

Consider the following problem which is a mixture of our simple one-dimensional diffusion and advection problems:

$$u_t + au_x = \epsilon u_{xx}, \quad \epsilon > 0. \tag{5.48}$$

Let us approximate it by central differences in space and a forward difference in time:

$$\frac{U^{n+1} - U^n}{\Delta t} + a\frac{\Delta_{0x}U^n}{\Delta x} = \epsilon\frac{\delta_x^2 U^n}{(\Delta x)^2}. \tag{5.49}$$

We now have two mesh ratios

$$\nu := a\Delta t/\Delta x, \quad \mu := \epsilon\Delta t/(\Delta x)^2 \tag{5.50}$$

and a Fourier analysis gives the amplification factor

$$\lambda(k) = 1 - i\nu \sin k\Delta x - 4\mu \sin^2 \tfrac{1}{2}k\Delta x, \tag{5.51a}$$

$$|\lambda|^2 = (1 - 4\mu s^2)^2 + 4\nu^2 s^2(1 - s^2), \tag{5.51b}$$

where as usual $s = \sin \tfrac{1}{2}k\Delta x$. Putting $s^2 = 1$ shows that $\mu \leq \tfrac{1}{2}$ is necessary for stability; then

$$\nu^2 = (a\Delta t/\Delta x)^2 = (a^2/\epsilon)\mu\Delta t$$

and this implies

$$|\lambda|^2 \leq 1 + \tfrac{1}{2}(a^2/\epsilon)\Delta t \tag{5.52}$$

so that the von Neumann condition is satisfied. As this is a scalar pure initial-value problem, and a and ϵ are constants this is sufficient for stability. However, if $\nu = 1$, $\mu = \frac{1}{4}$ and $s^2 = \frac{1}{2}$ we have $|\lambda|^2 = \frac{5}{4}$ giving very rapid growth; by contrast, the differential problem damps all Fourier modes.

In practice, then, for finite values of Δx and Δt, the von Neumann condition is too weak when the exponential growth that it allows is inappropriate to the problem. We therefore introduce the following stricter definition:

Definition 5.1 *A scheme is said to be practically (or strictly or strongly) stable if, when Fourier mode solutions of the differential problem satisfy*

$$|\widehat{u}(\mathbf{k}, t + \Delta t)| \leq e^{\alpha \Delta t}|\widehat{u}(\mathbf{k}, t)| \quad \forall \mathbf{k} \tag{5.53}$$

for some $\alpha \geq 0$, then the corresponding amplification factors for the difference scheme satisfy

$$|\lambda(\mathbf{k})| \leq e^{\alpha \Delta t} \tag{5.54}$$

for all $\mathbf{k}$ that correspond to discrete modes.

For the above example, we have $\alpha = 0$ and so require $|\lambda| \leq 1$. From (5.51b) we have

$$|\lambda|^2 = 1 - 4(2\mu - \nu^2)s^2 + 4(4\mu^2 - \nu^2)s^4. \tag{5.55}$$

By considering this expression as a positive quadratic function of s^2 on the interval $[0, 1]$, we obtain the conditions

$$|\lambda|^2 \leq 1 \quad \forall \mathbf{k} \quad \text{iff} \quad \nu^2 \leq 2\mu \leq 1. \tag{5.56}$$

These are very well known and often important restrictions because they can be very severe if ϵ is small: apart from the expected condition $\mu \leq \frac{1}{2}$, we can write the first inequality as

$$\frac{a.a\Delta t}{\epsilon} \equiv \frac{\nu^2}{\mu} \leq 2. \tag{5.57}$$

So it can be interpreted as placing a limit of 2 on a mesh Péclet number, in whose definition a is a velocity, $a\Delta t$ is interpreted as a mesh-length and ϵ is a diffusion or viscosity coefficient – see also the discussion in Section 2.15 where such restrictions arise from considering the application of a maximum principle.

Indeed, such is the practical importance of this criterion and resulting mesh restriction, the definition embodied in (5.53) and (5.54) is often used in the engineering field as the main definition of stability, the term being used without qualification. Note also that in Section 5.9 we refer to the property as *strong stability* when applied to problems which have no solution growth, corresponding to the concept of absolute stability in the discretisation of ODEs.

Using the same analysis as that given above, similar practical stability criteria can be found by combining any of the explicit schemes we have analysed for the diffusion equation with ones we have used for linear advection. Generally speaking the resultant condition will be a more severe combination of the corresponding results from the two parts. For example, if the upwind method (4.13) is used in (5.49) instead of the central difference we get

$$|\lambda|^2 \leq 1 \quad \forall k \quad \text{iff} \quad \nu^2 \leq \nu + 2\mu \leq 1 \tag{5.58}$$

compared with the conditions $0 \leq \nu \leq 1$ and $0 \leq \mu \leq \frac{1}{2}$, which are needed for the separate equations.

In carrying out these stability calculations, it is convenient to use general criteria for the roots of a polynomial to lie inside the closed unit disc. Miller[1] has called such polynomials von Neumann polynomials and gives criteria for mth degree polynomials in terms of those for related polynomials of degree $m-1$. We will just give the criteria for an arbitrary quadratic with complex coefficients, which generalise criteria we have devised for special cases in earlier chapters.

Lemma 5.2 *The roots of the polynomial $a\lambda^2 + 2b\lambda + c = 0$ with complex coefficients a, b, c satisfy the condition $|\lambda| \leq 1$ if and only if*

$$either \quad |c| < |a| \quad and \quad 2|\bar{a}b - \bar{b}c| \leq |a|^2 - |c|^2, \tag{5.59a}$$

$$or \quad |c| = |a|, \quad \bar{a}b = \bar{b}c \quad and \quad |b| \leq |a|. \tag{5.59b}$$

The proof is left as an exercise – see Exercise 3.

Finally, we refer again to the conditions given in Section 2.11 for the θ-method applied to the simple heat flow problem to be stable or satisfy a maximum principle. We saw that $\mu(1 - \theta) \leq \frac{1}{2}$ gave a maximum principle and ensured stability and no error growth in the maximum norm; on the other hand, $\mu(1 - 2\theta) \leq \frac{1}{2}$ ensured stability in the l_2 norm, and we have since seen that this also guarantees stability in the

---

[1] Miller, J. (1971), On the location of zeros of certain classes of polynomials with applications to numerical analysis, *J. Inst. Math. Appl.* **8**, 397–406.

sense of (5.28) in the maximum norm. However, we quoted a result in Section 2.11 to the effect that for no growth in the maximum norm the necessary and sufficient condition is $\mu(1 - \theta)^2 \leq \frac{1}{4}(2 - \theta)$. We can now note that this corresponds to practical stability in the maximum norm, as expressed by (2.97) with $K = 1$.

5.8 Modified equation analysis

This type of analysis was originally introduced as an alternative means of deriving stability conditions, but one has to be rather careful in using it for this purpose. It is now regarded as more useful in giving information about the general behaviour of a numerical method and the quality of the approximation it produces. The name we have used is due to Warming and Hyett.[1] However, similar ideas were introduced earlier by Yanenko and Shokin[2] and developed at length in Shokin (1983) under the name *method of differential approximation*. The general idea is to fit a smooth function through the mesh values $\{U_j^n\}$ and thence to find, to a given order in the mesh parameters, the differential equation that it satisfies. More recently, by analogy with techniques that have long been common in numerical linear algebra, the techniques have been called *backward error analysis*. We will illustrate the ideas by means of a few examples.

To develop a rigorous analysis it is probably best to extend the mesh data by means of Fourier expansion, but a simpler procedure is to use polynomial interpolation. The manipulations that are then needed follow closely those that we have used hitherto to estimate truncation errors. So let us begin with the upwind scheme applied to the linear advection equation as in Sections 4.2 and 4.3. We denote by $\tilde{U}(x,t)$ an interpolatory polynomial through the mesh values of U in the neighbourhood of the mesh point (x_j, t_n). Then when this is substituted in the difference equation (4.13) we can expand the two divided differences by (truncated) Taylor series to obtain, as in (4.24),

$$[\tilde{U}_t + \tfrac{1}{2}\Delta t\tilde{U}_{tt} + \cdots]_j^n + [a(\tilde{U}_x - \tfrac{1}{2}\Delta x\tilde{U}_{xx} + \cdots)]_j^n = 0. \qquad (5.60)$$

[1] Warming, R.F. and Hyett, B.J. (1974), The modified equation approach to the stability and accuracy of finite difference methods, *J. of Comput. Phys.* **14**, 159–79.

[2] Yanenko, N.N. and Shokin, Y.I. (1969), First differential approximation method and approximate viscosity of difference schemes, *Phys. of Fluids* **12**, Suppl. II, 28–33.

According to the order of the polynomial used for $\tilde{U}$, these expansions will be truncated at points that define a *modified equation* that it satisfies; but, of course, the degree of the polynomial is also closely related to the neighbourhood of the mesh point for which the approximation is valid. For example, at one extreme we can fit a linear polynomial through the three values appearing in the difference scheme and this exactly satisfies the original linear advection equation; but this gives us no global information about the approximation provided by the difference scheme.

In general, then, we shall expand to a higher order that we will not specify. We can drop the superscripts and subscripts to obtain

$$\tilde{U}_t + a\tilde{U}_x = \tfrac{1}{2}[-\Delta t \tilde{U}_{tt} + a\Delta x \tilde{U}_{xx}] + \cdots . \tag{5.61}$$

This is still not very useful because of the higher order time derivatives appearing on the right-hand side. One way of dealing with this is to operate on (5.61) with the combination $\partial_t - a\partial_x$ which, if we assume for the moment that a is a positive constant, will give an expression for $\tilde{U}_{tt}$ in terms of $\tilde{U}_{xx}$. Substitution into the right-hand side of (5.61) then gives

$$\tilde{U}_t + a\tilde{U}_x = \tfrac{1}{2}[-a^2\Delta t + a\Delta x]\tilde{U}_{xx} + \cdots . \tag{5.62}$$

We can deduce from this that a choice of $a\Delta t = \Delta x$ will mean that $\tilde{U}$ satisfies an equation that differs from the target advection equation by terms that are at least of second order in the mesh parameters; also we see that a choice with $a\Delta t < \Delta x$ will mean that its equation has a first order perturbation giving damping through a diffusion term; but, more importantly, if $a\Delta t > \Delta x$ the diffusion term will have a negative coefficient and *this will lead to severe error growth, and corresponds to instability.*

A more powerful procedure than that applied above is to make use of finite difference operator calculus – see, for instance, Chapter 5 of Hildebrand (1956). Thus, by recognising that the coefficients in a Taylor series expansion are just those of an exponential series, we can write the forward time difference in operator form as

$$\Delta_{+t} = e^{\Delta t \partial_t} - 1; \tag{5.63}$$

this can be formally inverted to give the inverse relation

$$\partial_t = (1/\Delta t)\ln(1 + \Delta_{+t}). \tag{5.64}$$

Then from this we derive an expansion for ∂_t in terms of $\mathcal{D}_{+t} := \Delta_{+t}/\Delta t$,

$$\partial_t = \mathcal{D}_{+t} - \tfrac{1}{2}\Delta t\mathcal{D}_{+t}^2 + \tfrac{1}{3}(\Delta t)^2\mathcal{D}_{+t}^3 - \tfrac{1}{4}(\Delta t)^3\mathcal{D}_{+t}^4 + \cdots . \tag{5.65}$$

If a difference scheme gives an expression for $\mathcal{D}_{+t}$ in terms of spatial derivatives, this expansion will lead directly to a modified equation expansion of the desired form.

So let us consider the central difference in space and forward difference in time scheme (5.49), applied to the convection–diffusion problem (5.48). We expand the spatial difference operators to get

$$\mathcal{D}_{+t}U = \{-a[\partial_x + \tfrac{1}{6}(\Delta x)^2\partial_x^3 + \cdots] + \epsilon[\partial_x^2 + \tfrac{1}{12}(\Delta x)^2\partial_x^4 + \cdots]\}\tilde{U}. \tag{5.66}$$

Then we substitute this into the expansion (5.65) and collect terms. For later use we will expand up to terms in the fourth spatial derivative, and obtain

$$\begin{aligned}
\partial_t\hat{U} = \{&[(-a\partial_x + \epsilon\partial_x^2) + (\Delta x)^2(-\tfrac{1}{6}a\partial_x^3 + \tfrac{1}{12}\epsilon\partial_x^4 - \cdots)] \\
&- \tfrac{1}{2}\Delta t[a^2\partial_x^2 - 2a\epsilon\partial_x^3 + (\epsilon^2 + \tfrac{1}{3}a^2(\Delta x)^2)\partial_x^4 + \cdots] \\
&+ \tfrac{1}{3}(\Delta t)^2[-a^3\partial_x^3 + 3a^2\epsilon\partial_x^4 + \cdots] \\
&- \tfrac{1}{4}(\Delta t)^3[a^4\partial_x^4 + \cdots] + \cdots \}\tilde{U},
\end{aligned} \tag{5.67}$$

which can be written as the modified equation

$$\tilde{U}_t + a\tilde{U}_x = [\epsilon - \tfrac{1}{2}a^2\Delta t]\tilde{U}_{xx} - \tfrac{1}{6}[a(\Delta x)^2 - 6a\epsilon\Delta t + 2a^3(\Delta t)^2]\tilde{U}_{xxx} + \cdots . \tag{5.68}$$

We see immediately that for all choices of the time step the diffusion coefficient is reduced from that in the original equation; and it is non-negative only for

$$\tfrac{1}{2}a^2\Delta t \le \epsilon, \quad \text{or} \quad \frac{a.(a\Delta t)}{\epsilon} \le 2.$$

This is the stability limit corresponding to the requirement that the mesh Péclet number be less than two, where the length scale is given by $a\Delta t$; it is equivalent to the practical stability condition contained in (5.56) and (5.57) that we have $\nu^2 \le 2\mu$, in terms of the mesh ratios defined in (5.50). Note that the next term in (5.68) allows the amount of dispersion in the scheme to be estimated.

We should remark at this point that the interpretation of a modified equation, such as those obtained above, owes a great deal to Fourier analysis. Thus the effect of a spatial derivative of order m on a typical Fourier mode e^{ikx} is multiplication by a factor $(ik)^m$. Hence odd order derivatives change the phase of a mode (leading to dispersion), while only even order derivatives change its magnitude and can lead to instability. If the coefficient of the second order derivative is negative, as we have seen can happen in the example just considered, one can deduce instability because this term will dominate all higher order terms at small values of $k\Delta x$. If this coefficient is zero we will need to go to higher order terms, which we shall do below.

We have given the above analysis in some detail because it can be used to compare the behaviour of any explicit three point approximation to the convection–diffusion problem. For consistency the first order difference should be $a\Delta_{0x}/\Delta x$, but the second order difference can have a different coefficient. So suppose we change it from ϵ to ϵ'. Then we can deduce, from (5.67) with this change, that the diffusion term is correctly modelled if we set $\epsilon' - \frac{1}{2}a^2\Delta t = \epsilon$; that is, $\epsilon'\Delta t/(\Delta x)^2 = \epsilon\Delta t/(\Delta x)^2 + \frac{1}{2}(a\Delta t/\Delta x)^2$ which corresponds exactly to what is given by the Lax–Wendroff scheme. The dominant error is then dispersive, with the coefficient of $\tilde{U}_{xxx}$ given by

$$-\tfrac{1}{6}a(\Delta x)^2 + a\Delta t(\epsilon + \tfrac{1}{2}a^2\Delta t) - \tfrac{1}{3}a^3(\Delta t)^2$$
$$= -\tfrac{1}{6}a(\Delta x)^2[1 - (a\Delta t/\Delta x)^2] + a\epsilon\Delta t. \tag{5.69}$$

We deduce from this that, for pure advection, the dispersive error of the Lax–Wendroff method is always negative for any stable mesh ratio – the expression given above being consistent in form and magnitude with the predominant phase lag error for this scheme shown in Section 4.5.

To consider the stability of the Lax–Wendroff method we need to consider terms in the expansion (5.67) involving the fourth spatial derivative. We consider only the case $\epsilon = 0$, i.e., for pure advection; then with the CFL number given by $\nu = a\Delta t/\Delta x$, it is readily found that the modified equation for this method is given by

$$\tilde{U}_t + a\tilde{U}_x = -\tfrac{1}{6}a(\Delta x)^2[1 - \nu^2]\tilde{U}_{xxx} - \tfrac{1}{8}a^2\Delta t(\Delta x)^2[1 - \nu^2]\tilde{U}_{xxxx} + \cdots. \tag{5.70}$$

Thus the coefficient of the fourth derivative becomes positive when $|\nu| > 1$, corresponding to instability.

This is a good example in which the second order coefficient is zero but the coefficient of the fourth order derivative can become positive, when it will dominate at small values of $k\Delta x$ and we can deduce instability. However, if both coefficients are positive the situation is much more complicated: we would expect that the second order term would dominate and hence imply stability; but we would have to consider the whole range of Fourier modes that the mesh can carry, which we recall is such that $|k\Delta x| \leq \pi$, and all terms in the modified equation expansion.

We have therefore concentrated in the above considerations of stability on the pure advection case: for it is easily seen that these modified equation arguments break down even for the simple heat flow problem. For example, suppose we set $a = 0$ in (5.67) and try to deduce the familiar stability limit. We will find ourselves in the situation outlined in the previous paragraph and not be able to make a sensible deduction. The underlying reason for distinguishing between the two cases is clear from our earlier Fourier analysis. The instability of many approximations to the advection equation shows itself at low frequencies, with plots of the amplification factor such as in Fig. 4.18 moving outside the unit circle as $k\Delta x$ increases from zero; and for such modes an expansion such as (5.67) in increasing order of the spatial derivatives makes good sense. But, as we saw in Sections 2.7 and 2.10, the amplification factor for a typical approximation to the heat flow problem is real and instability occurs first for the most oscillatory mode $k\Delta x = \pi$ when the amplification factor becomes less than -1. Then an expansion like (5.67) is of no use.

There is a solution to this dilemma, however. For each of the modes in the range $|k\Delta x| \leq \pi$, we can write $k\Delta x = \pi - k'\Delta x$, where $|k'\Delta x|$ is small for the most oscillatory modes; then expansions in powers of this quantity will correspond to expansions in spatial derivatives of the amplitudes of the oscillatory modes. Equivalent to splitting the range of the Fourier modes, let us write

$$U_j^n = (U^s)_j^n + (-1)^{j+n}(U^o)_j^n, \tag{5.71}$$

so that U is split into a smooth part U^s and an oscillatory part U^o, and try to find a modified equation that describes the behaviour of the latter. Note that we have also taken out a factor $(-1)^n$ in the time variation, which corresponds to our expectation that these modes become unstable with an amplification factor near -1.

Let us therefore consider now the same scheme as above in (5.66) but with $a = 0$ to give the pure diffusion case. For the oscillatory modes we take out the common factor $(-1)^{j+n}$ and obtain

$$\begin{aligned}
-(U^o)_j^{n+1} &= (U^o)_j^n + \mu[-(U^o)_{j-1}^n - 2(U^o)_j^n - (U^o)_{j+1}^n] \\
&= (1 - 2\mu)(U^o)_j^n \\
&\quad - 2\mu[1 + \tfrac{1}{2}(\Delta x)^2 \partial_x^2 + \tfrac{1}{24}(\Delta x)^4 \partial_x^4 + \cdots](U^o)_j^n,
\end{aligned} \tag{5.72}$$

where the expansion in square brackets results from the average of the terms at $j \pm 1$. Hence, after collecting together these terms, we obtain for the forward divided difference

$$\mathcal{D}_{+t}U^o = \{2(\Delta t)^{-1}(2\mu - 1) + \epsilon[\partial_x^2 + \tfrac{1}{12}(\Delta x)^2\partial_x^4 + \cdots]\}U^o. \quad (5.73)$$

We see immediately that we obtain exponential growth when $\mu > \tfrac{1}{2}$.

The same technique can help to understand some of the erratic behaviour in Fig. 2.7. Writing the equivalent of (5.73) but for the Crank–Nicolson method we easily find that

$$D_{+t}U^o = \frac{-2}{(2\mu + 1)\Delta t}U^o + O((\Delta x)^2). \quad (5.74)$$

This is clearly stable, giving an exponentially decreasing term for any positive value of μ. As Δt goes to zero the exponential factor is vanishingly small except for extremely small t. However, on a coarse mesh the situation is different: for example, when $J = 10, \mu = 10$ we see that $\Delta t = 1/10$, so the exponential factor is $\exp(-(20/21)t)$, which is significant over the whole range $0.1 \leq t \leq 1$. This suggests that the oscillatory term has a significant effect on the maximum error in this range. As the mesh is refined the effect rapidly gets smaller, and this analysis suggests that the oscillatory terms become negligible when J reaches about 50.

We conclude this section by applying this analysis to the box scheme of Section 4.8, where the $(-1)^{j+n}$ mode is the notorious chequerboard mode which is a spurious solution of the difference scheme controlled only by the boundary conditions. It is convenient to use the averaging operator

$$\mu_x U_{j+1/2} := \tfrac{1}{2}(U_j + U_{j+1}), \quad (5.75)$$

and similarly μ_t, in terms of which we can write the box scheme for the linear (constant coefficient) advection equation as

$$\mu_x \delta_t U_{j+1/2}^{n+1/2} + (a\Delta t/\Delta x)\mu_t \delta_x U_{j+1/2}^{n+1/2} = 0. \quad (5.76)$$

We could obtain the modified equation given below by expanding all four mesh values in Taylor series about their central point, and then substituting space derivatives for higher order time derivatives in the manner used to derive (5.62) from (5.61). However, a more illuminating and sophisticated approach is to proceed as follows: we begin by operating on the box scheme (5.76) with the inverse of the two averaging operators, and introduce the difference operators

$$\mathcal{D}_x = \frac{\mu_x^{-1}\delta_x}{\Delta x}, \quad \mathcal{D}_t = \frac{\mu_t^{-1}\delta_t}{\Delta t}. \quad (5.77)$$

Then we can write the scheme in the compact form $(\mathcal{D}_t + a\mathcal{D}_x)U = 0$. Though this step is a formal procedure, and not one to be executed in practice, it demonstrates the consistent way in which the box scheme replaces differential operators by difference analogues, and thence yields a relation that is a direct analogue of the differential equation. Moreover, it points the way to use operator calculus to derive the modified equation in a straightforward and very general way.

The first step is to carry out a manipulation similar to that in (5.63) and (5.64) to obtain the difference operator relation

$$\mathcal{D}_x = (\tfrac{1}{2}\Delta x)^{-1} \tanh(\tfrac{1}{2}\Delta x \partial_x),$$

from which we can deduce that

$$\mathcal{D}_x = [1 - \tfrac{1}{12}(\Delta x)^2 \partial_x^2 + \cdots]\partial_x. \tag{5.78}$$

Then we carry out a similar manipulation with the time difference, but invert it to obtain

$$\partial_t = [1 + \tfrac{1}{12}(\Delta t)^2 \mathcal{D}_t^2 + \cdots]\mathcal{D}_t. \tag{5.79}$$

Application of the two expansions (5.78, 5.79) immediately yields the modified equation for the scheme

$$\tilde{U}_t + a\tilde{U}_x = \tfrac{1}{12}a(\Delta x)^2(1 - \nu^2)\tilde{U}_{xxx} + \cdots, \tag{5.80}$$

where we have introduced the CFL number ν.

All of the extra terms in this expansion will involve odd orders of derivatives, consistent with the unconditional stability of the scheme; and the dispersion coefficient changes sign according to whether we have $\nu < 1$ (giving a phase advance) or $\nu > 1$ (giving retardation). Indeed, the expansion (5.80) corresponds exactly to the expansion for the phase of the amplification factor given in (4.85). However, the modified equation can be generalised to cover a variable advection speed as in the example giving Fig. 4.13, or indeed to a nonlinear problem.

The modified equation (5.80) applies to only the smooth component of the flow. But Fig. 4.13 shows that discontinuous initial data produce violent fluctuations in the solution. So let us now derive the corresponding modified equation for the oscillatory part of the decomposition (5.71). For the difference operators we are using it is easy to see that for the oscillatory part of the decomposition (5.71) we have

$$\delta_x U_{j+1/2} = (-1)^{j+1} 2\mu_x U_{j+1/2}^o,$$

$$\mu_x U_{j+1/2} = (-1)^{j+1} \tfrac{1}{2}\delta_x U_{j+1/2}^o; \tag{5.81}$$

that is, differences are turned into averages and averages into differences. Hence the equation for U^o obtained from (5.76) is just

$$\mu_t \delta_x (U^o)_{j+1/2}^{n+1/2} + (a\Delta t/\Delta x)\mu_x \delta_t (U^o)_{j+1/2}^{n+1/2} = 0. \qquad (5.82)$$

By comparing the two equations we can immediately write down the modified equation for the oscillatory component: it is easy to see that U^o satisfies a box scheme for an advection equation with a replaced by $1/\nu_2$ where $\nu_2 = a(\Delta t/\Delta x)^2$ so we obtain

$$\tilde{U}_t^o + (1/\nu_2)\tilde{U}_x^o = \tfrac{1}{12}(1/\nu_2)(\Delta x)^2(1 - \nu^{-2})\tilde{U}_{xxx}^o + \cdots . \qquad (5.83)$$

Thus these modes are again undamped but generally they are advected at quite a different speed from a and one that depends on the mesh ratio. As we saw in Section 4.8, this speed can also be deduced from the group velocity of these modes, $C_h = a/\nu^2 \equiv 1/\nu_2$.

As also pointed out in that section, the chequerboard mode is usually controlled by a weighted time-average. So we end by looking at the effect of this on the modified equations. Let us define the averaging operator

$$\theta_t U^{n+1/2} := \theta U^{n+1} + (1 - \theta)U^n, \qquad (5.84)$$

so that we have $\theta_t = \mu_t + (\theta - \tfrac{1}{2})\delta_t$. Then we introduce the operator

$$\mathcal{M}_t := \mu_t^{-1}\theta_t = 1 + (\theta - \tfrac{1}{2})\Delta t \mathcal{D}_t, \qquad (5.85)$$

in terms of which the weighted box scheme becomes

$$\left[\mathcal{D}_t + a\mathcal{M}_t \mathcal{D}_x\right]U_{j+1/2}^{n+1/2} = 0. \qquad (5.86)$$

Writing $\gamma = (\theta - \tfrac{1}{2})\Delta t$, and solving for the time difference, we obtain the expansion

$$\mathcal{D}_t U = (1 + \gamma a\mathcal{D}_x)^{-1}(-a\mathcal{D}_x)U$$
$$= \left[-a\mathcal{D}_x + \gamma a^2 \mathcal{D}_x^2 - \gamma^2 a^3 \mathcal{D}_x^3 + \cdots\right]U. \qquad (5.87)$$

Then substituting from (5.78, 5.79) we obtain for the modified equation

$$\tilde{U}_t + a\tilde{U}_x = \gamma a^2 \tilde{U}_{xx}$$
$$+ a(\Delta x)^2\left[\tfrac{1}{12}(1 - \nu^2) - \gamma^2 a^2\right]\tilde{U}_{xxx} + \cdots . \qquad (5.88)$$

Thus taking $\theta < \tfrac{1}{2}$ gives $\gamma < 0$ and leads to instability; but taking $\theta > \tfrac{1}{2}$ introduces some damping, which is usually limited by setting $\theta = \tfrac{1}{2} + O(\Delta t)$.

Consider now the oscillating modes. Instead of (5.86) we obtain from (5.81)

$$\left\{ \mathcal{D}_x + a(\Delta t/\Delta x)^2 \left[\mathcal{D}_t + 4(\Delta t)^{-1}(\theta - \tfrac{1}{2})\right] \right\} U^o = 0, \tag{5.89}$$

which yields the modified equation

$$\tilde{U}_t^o + (1/\nu_2)\tilde{U}_x^o = -4(\Delta t)^{-1}(\theta - \tfrac{1}{2})\tilde{U}^o + \cdots . \tag{5.90}$$

Hence even with $\theta = \tfrac{1}{2} + O(\Delta t)$ there is exponential damping of the chequerboard mode for $\theta > \tfrac{1}{2}$.

5.9 Conservation laws and the energy method of analysis

We have already seen in Sections 2.7 and 4.11 how conservation laws for $\int u \, dx$ can be helpful in the choice of boundary conditions. Now we shall consider how the use of similar laws for $\int |u|^2 \, dx$ can lead to useful methods for establishing stability. Such arguments have their origin and draw their motivation from a number of sources. Undoubtedly the most important is the role of energy inequalities in establishing the well-posedness of large classes of PDEs written in the form (5.1a). Suppose that by taking the inner product of both sides with u, integrating over Ω and exploiting Green's identities, one can show that

$$(u, Lu)_2 \equiv \int_\Omega u L(u) \, d\Omega \leq K \int_\Omega |u|^2 \, d\Omega \equiv K\|u\|_2^2 \tag{5.91}$$

for certain boundary conditions. Then this will establish the well-posedness of the differential problem in the L_2 norm. Because this norm sometimes represents the physical energy of the system such approaches are generally called *energy methods*, whether applied to the differential problem or its finite difference approximation.

As a simple example we return to the model problem introduced at the beginning of Chapter 2, the heat equation in one dimension with homogeneous Dirichlet boundary conditions:

$$u_t = u_{xx}, \quad 0 < x < 1, \ t > 0, \tag{5.92a}$$
$$u(0,t) = u(1,t) = 0, \quad t > 0, \tag{5.92b}$$
$$u(x,0) = u^0(x), \quad 0 < x < 1. \tag{5.92c}$$

From (5.92) it immediately follows that $\left(\frac{1}{2}u^2\right)_t = u\,u_{xx}$, so that after integration by parts

$$\frac{\partial}{\partial t}\left(\int_0^1 \tfrac{1}{2}u^2\,\mathrm{d}x\right) = \int_0^1 u\,u_{xx}\,\mathrm{d}x = \left[u\,u_x\right]_0^1 - \int_0^1 (u_x)^2\,\mathrm{d}x \quad (5.93)$$

$$= -\int_0^1 (u_x)^2\,\mathrm{d}x \leq 0, \quad (5.94)$$

where we have also used the boundary conditions (5.92b). We have thus shown that the L_2 norm of u is decreasing, and therefore bounded by its initial value. Note that this does not correspond to the physical energy of the system in this case.

The main tool in this analysis was integration by parts. To obtain a similar result in the discrete case we shall need corresponding *summation by parts* formulae which we shall give below. But first we note that such analysis in the discrete case has its direct motivation from the Kreiss Matrix Theorem, already referred to in Section 5.6. One of the statements in that theorem that is shown to be equivalent to the power-boundedness of the amplification matrices G, as expressed in (5.46), is that there exists an energy norm, $|v|_H^2 := v^*Hv$, for which we have $|G|_H \leq 1$; that is, there exists a uniformly bounded and positive definite hermitian matrix H for which we have

$$G^*HG \leq H. \quad (5.95)$$

So in this norm at each time step there is no growth. Unfortunately, we do not know in general how to construct or choose these matrices H, or the operators that they correspond to when we move from the Fourier space to real space; and it is these operators in real space that we need because a key objective of the theorem is to lead to the extension of the stability results obtained by Fourier analysis to problems with variable coefficients, nonperiodic boundary conditions, etc. On the other hand, in many important cases we can construct such operators from judicious combinations of difference operators, and in this section we will illustrate how this is done by means of a number of examples. We can then establish the stability of the scheme being considered, but we cannot demonstrate instabilities; so the approach yields sufficient stability conditions to complement the necessary conditions given by the von Neumann analysis.

Before embarking on this, however, we state a recent result[1] that enables a significant class of methods to be proved stable in a very straightforward way; the theorem in fact gives conditions for *strong stability*, which is equivalent to our definition of practical stability as it applies directly only to problems that have no solution growth.

Theorem 5.3 *Suppose that a well-posed problem in the form (5.1a) is approximated by (repeatedly) applying a spatial difference operator L_Δ and incorporating it into an expansion to give*

$$U^{n+1} = \sum_{i=0}^{s} \frac{(\Delta t L_\Delta)^i}{i!} U^n, \qquad (5.96)$$

which we will refer to as a Runge–Kutta time-stepping procedure. Suppose further that the operator L_Δ is coercive in the sense that there is a positive constant η such that

$$(L_\Delta U, U) \le -\eta \|L_\Delta U\|^2 \qquad (5.97)$$

for all mesh functions U. Then the scheme is strongly stable for $s = 1, 2, 3,$ or 4 if the time step satisfies the condition

$$\Delta t \le 2\eta. \qquad (5.98)$$

Proof For a method with no repetitive application of the operator, i.e. with $s = 1$, in which the operator is coercive in the sense of (5.97), we have

$$
\begin{aligned}
\|U^{n+1}\|^2 &= \|U^n + \Delta t L_\Delta U^n\|^2 \\
&= \|U^n\|^2 + 2\Delta t (U^n, L_\Delta U^n) + (\Delta t)^2 \|L_\Delta U^n\|^2 \\
&\le \|U^n\|^2 + \Delta t (\Delta t - 2\eta) \|L_\Delta U^n\|^2. \qquad (5.99)
\end{aligned}
$$

So, if (5.98) is satisfied, the norm of the approximate solution is non-increasing. For $s = 2$ it is easily checked that we can write

$$I + \Delta t L_\Delta + \tfrac{1}{2}(\Delta t L_\Delta)^2 = \tfrac{1}{2}I + \tfrac{1}{2}(I + \Delta t L_\Delta)^2,$$

for $s = 3$ we can similarly check that we can write

$$\sum_{i=0}^{3} \frac{(\Delta t L_\Delta)^i}{i!} = \tfrac{1}{3}I + \tfrac{1}{2}(I + \Delta t L_\Delta) + \tfrac{1}{6}(I + \Delta t L_\Delta)^3,$$

---

[1] Levy, D. and Tadmor, E. (1998), From semi-discrete to fully discrete: stability of Runge–Kutta schemes by the energy method, *SIAM Rev.* **40**, 40–73; and Gottlieb, S., Shu, C.-W. and Tadmor, E. (2001), Strong stability-preserving high-order time discretisation methods, *SIAM Rev.* **43**, 89–112.

and for $s = 4$ we have

$$\sum_{i=0}^{4} \frac{(\Delta t L_\Delta)^i}{i!} = \tfrac{3}{8} I + \tfrac{1}{3}(I + \Delta t L_\Delta) + \tfrac{1}{4}(I + \Delta t L_\Delta)^2 + \tfrac{1}{24}(I + \Delta t L_\Delta)^4.$$

Now we have already shown that if (5.98) is satisfied then $\|I + \Delta t L_\Delta\| \le 1$. So all powers of this operator are similarly bounded and, since the coefficients in each of these expansions sum to unity, then all three operator expansions are bounded by unity. $\qquad\square$

Before we make any use of this theorem some comments are in order. Its motivation comes from the use of *semi-discrete methods*; that is, those in which the spatial discretisation is carried out first so as to lead to a large system of ODEs to which, for instance, Runge–Kutta time-stepping methods can be applied. In order for this to be done with $s > 1$ one would clearly need to pay attention to the application of appropriate boundary conditions; and to apply the ideas to nonlinear problems, further care would be needed. But here we will confine ourselves to our standard linear problems, and apply the theorem only to the case $s = 1$. Yet it still remains a useful framework in which to establish the stability of several methods by energy analysis.

The real inner product and norm in the theorem can be quite general, but our purpose here is to apply the result using the discrete l_2 norm. In general this will be of the form in (5.5) but we will confine our examples to one dimension and a uniform mesh: thus the l_2 inner product of two real vectors U and V when $J_\Omega = \{1, 2, \ldots, J - 1\}$ is given by

$$\langle U, V \rangle_2 = \Delta x \sum_{1}^{J-1} U_j V_j, \tag{5.100}$$

for which we have $\|U\|_2 = \langle U, U \rangle_2^{1/2}$. On this mesh the summation by parts formulae are given in the following lemma.

Lemma 5.3 *For any sequences of numbers $\{V_j\}$ and $\{W_j\}$*

$$\sum_{1}^{J-1} [V_j(W_{j+1} - W_j) + W_j(V_j - V_{j-1})] = V_{J-1} W_J - V_0 W_1 \tag{5.101}$$

and

$$\sum_{1}^{J-1} \left[V_j \left(W_{j+1} - W_{j-1}\right) + W_j(V_{j+1} - V_{j-1})\right]$$
$$= \left(V_{J-1}W_J + W_{J-1}V_J\right) - \left(V_0W_1 + W_0V_1\right). \quad (5.102)$$

The proof of (5.101) involves straightforward algebraic manipulation and is left as an exercise. The second formula (5.102) is obtained by interchanging V and W in (5.101) and then adding the two results together.

In terms of difference operators we can write (5.101) and (5.102) as

$$\langle V, \Delta_{+x}W\rangle_2 + \langle W, \Delta_{-x}V\rangle_2 = \Delta x \left(V_{J-1}W_J - V_0W_1\right) \quad (5.103)$$

and

$$\langle V, \Delta_{0x}W\rangle_2 + \langle W, \Delta_{0x}V\rangle_2 = \tfrac{1}{2}\Delta x \left[(V_{J-1}W_J + W_{J-1}V_J) \right.$$
$$\left. -(V_0W_1 + W_0V_1)\right]. \quad (5.104)$$

By replacing W with $\Delta_{-x}W$ in the first formula we also get

$$\langle V, \delta_x^2 W\rangle_2 + \langle \Delta_{-x}W, \Delta_{-x}V\rangle_2 = \Delta x \left[V_{J-1}(W_J - W_{J-1}) - V_0(W_1 - W_0)\right]; \quad (5.105)$$

and hence if we have $V = W$ and $V_0 = V_J = 0$ we can write

$$\langle V, \delta_x^2 V\rangle_2 = -\|\delta_x V\|_2^2 := -\Delta x \sum_{j=1}^{J} |\delta_x V_{j-1/2}|^2, \quad (5.106)$$

where it should be noted that now the sum on the right is over all of the J cells.

Example 1 We consider first the use of the explicit method to solve (5.92). This can be written

$$U_j^{n+1} = U_j^n + \mu \delta_x^2 U_j^n \quad (5.107)$$

where $\mu = \Delta t/(\Delta x)^2$. Now we apply Theorem 5.3 to this with $L_\Delta = (\Delta x)^{-2}\delta_x^2$. From (5.106) we have

$$(\Delta x)^2(L_\Delta U, U) = \langle \delta_x^2 U, U\rangle = -\|\delta_x U\|^2, \quad (5.108)$$

where we have omitted the subscript on the norm $\| \cdot \|_2$ and inner product $\langle \cdot, \cdot \rangle_2$ (as we shall in the rest of this section); we can also use the Cauchy–Schwarz inequality to obtain

$$(\Delta x)^4 \| L_\Delta U \|^2 = \| \delta_x^2 U \|^2 = \Delta x \sum_{j=1}^{J-1} \left(\delta_x U_{j+1/2} - \delta_x U_{j-1/2} \right)^2$$

$$\leq 2 \Delta x \sum_{j=1}^{J-1} \left[(\delta_x U_{j+1/2})^2 + (\delta_x U_{j-1/2})^2 \right]$$

$$\leq 4 \| \delta_x U \|^2. \tag{5.109}$$

Comparing (5.108) with (5.109) it is clear that we can apply Theorem 5.3 with $\eta = \frac{1}{4}(\Delta x)^2$ and deduce the familiar result that we have stability for $\Delta t \leq \frac{1}{2}(\Delta x)^2$.

Example 2 Next we consider the implicit θ-method for the same problem, which is not covered by this theorem. So we proceed as follows: writing the equation as

$$U_j^n - U_j^{n-1} = \mu \left[\theta \delta_x^2 U_j^n + (1 - \theta) \delta_x^2 U_j^{n-1} \right], \tag{5.110}$$

we multiply both sides by the combination $U_j^n + U_j^{n-1}$ and sum over the values of j for which it holds, namely $j = 1, 2, \ldots, J - 1$. Thus we obtain

$$\| U^n \|^2 - \| U^{n-1} \|^2 = \mu \langle U^n + U^{n-1}, \delta_x^2 [\theta U^n + (1 - \theta) U^{n-1}] \rangle.$$

We can now apply the summation by parts formula (5.105) to the right-hand side, and follow that with an application of the Cauchy–Schwarz inequality in the form $-ab \leq \frac{1}{2}(a^2 + b^2)$, to get

$$\| U^n \|^2 - \| U^{n-1} \|^2 = -\mu \langle \delta_x [U^n + U^{n-1}], \delta_x [\theta U^n + (1 - \theta) U^{n-1}] \rangle$$

$$= -\mu \left\{ \theta \| \delta_x U^n \|^2 + (1 - \theta) \| \delta_x U^{n-1} \|^2 + \langle \delta_x U^n, \delta_x U^{n-1} \rangle \right\}$$

$$\leq -\mu \left\{ (\theta - \tfrac{1}{2}) \| \delta_x U^n \|^2 + (\tfrac{1}{2} - \theta) \| \delta_x U^{n-1} \|^2 \right\}. \tag{5.111}$$

Note that the coefficients in this final expression are equal and opposite. So if we now apply the same formula to all preceding steps, and add all the formulae, there will be cancellation of all the intermediate quantities. It is easy to see that the result is

$$\| U^n \|^2 - \| U^0 \|^2 \leq -\mu(\theta - \tfrac{1}{2}) \left\{ \| \delta_x U^n \|^2 - \| \delta_x U^0 \|^2 \right\}.$$

More meaningfully, we write this as

$$\| U^n \|^2 + \mu(\theta - \tfrac{1}{2}) \| \delta_x U^n \|^2 \leq \| U^0 \|^2 + \mu(\theta - \tfrac{1}{2}) \| \delta_x U^0 \|^2, \tag{5.112}$$

which will imply stability if these two expressions correspond to norms. Indeed, we have stability in the L_2 norm if in each case the first term dominates the second. Now in (5.109) we have already used the Cauchy–Schwarz inequality to obtain a bound for the norm of $\delta_x^2 U$ that we now need for $\delta_x U$; namely, from the argument there we deduce the useful inequality

$$\|\delta_x U\|^2 \leq 4\|U\|^2. \tag{5.113}$$

Applying this to (5.112) we deduce that $\mu(1 - 2\theta) < \frac{1}{2}$ gives stability, a result that is almost equivalent to that obtained in Section 2.10; we merely fail by this means to establish stability when we have equality in this relation, but this only affects cases with $\theta < \frac{1}{2}$. We note too that this is an example of our being able to construct explicitly the norm guaranteed by the Kreiss Matrix Theorem, because from (5.111) we see that if the stability condition is satisfied this norm is nonincreasing at each step.

Example 3 In this and the next example we will illustrate how an energy analysis can deal with more complicated problems, specifically those with variable coefficients and those needing the imposition of numerical boundary conditions. We consider the following problem,

$$u_t + a(x)u_x = 0, \quad 0 < x \leq 1, \quad t > 0, \tag{5.114a}$$

$$u(x,0) = u^0(x), \quad 0 \leq x \leq 1, \tag{5.114b}$$

$$u(0,t) = 0, \quad t > 0. \tag{5.114c}$$

We shall assume that $a(x)$ is nonnegative and bounded, and also satisfies a Lipschitz condition:

$$0 \leq a(x) \leq A, \quad |a(x) - a(y)| \leq K_L|x - y|. \tag{5.115}$$

Note that we have imposed a homogeneous Dirichlet boundary condition on the left, consistent with the characteristics pointing from left to right.

In this example we consider using the first order upwind scheme, to which we can apply a very direct argument as in Theorem 5.3. With $L_\Delta = -a(\Delta x)^{-1}\Delta_{-x}$, we have

$$\langle L_\Delta U, U \rangle = -\sum_{j=1}^{J-1} a_j(U_j - U_{j-1})U_j$$

$$= -\sum_{j=1}^{J-1} a_j U_j^2 + \sum_{j=1}^{J-1} a_j U_j U_{j-1}; \tag{5.116}$$

while we also have

$$\Delta x \|L_\Delta U\|^2 = \sum_{j=1}^{J-1} a_j^2 (U_j - U_{j-1})^2 \le A \sum_{j=1}^{J-1} a_j (U_j - U_{j-1})^2$$

$$= A \left\{ \sum_{j=1}^{J-1} a_j U_j^2 - 2 \sum_{j=1}^{J-1} a_j U_j U_{j-1} + \sum_{j=1}^{J-1} a_j U_{j-1}^2 \right\}. \quad (5.117)$$

Hence we can deduce that

$$2\langle L_\Delta U, U \rangle + A^{-1} \Delta x \|L_\Delta U\|^2 \le - \sum_{j=1}^{J-1} a_j (U_j^2 - U_{j-1}^2)$$

$$= - \sum_{j=1}^{J-2} (a_j - a_{j+1}) U_j^2 - a_{J-1} U_{J-1}^2$$

$$\le K_L \|U\|^2, \quad (5.118)$$

in which we have used the left-hand boundary condition for U and the Lipschitz condition on a. Hence it follows that if the stability condition $A\Delta t \le \Delta x$ is satisfied then

$$2\Delta t \langle L_\Delta U, U \rangle + \Delta t^2 \|L_\Delta U\|^2 \le \Delta t \left[2\langle L_\Delta U, U \rangle + A^{-1} \Delta x \|L_\Delta U\|^2 \right]$$

$$\le K_L \Delta t \|U\|^2. \quad (5.119)$$

We then conclude that

$$\|(I + \Delta t L_\Delta) U\|^2 \le (1 + K_L \Delta t) \|U\|^2, \quad (5.120)$$

which establishes the stability of the scheme. This example typifies the way in which the energy method can be used to show that the presence of variable coefficients does not destroy the stability of a method – though it may introduce some error growth.

Example 4 Now we consider the same advection problem (5.114), but apply the leap-frog method and for simplicity assume that a is a positive constant. The method may be written

$$U_j^{n+1} - U_j^{n-1} = -2\nu \Delta_{0x} U_j^n, \quad j \in J_\Omega \equiv \{1, 2, \dots, J-1\}, \quad (5.121)$$

where $\nu = a\Delta t/\Delta x$. It gives an explicit formula for U_j^{n+1}, but only at the internal points. The value of U_0^{n+1} is given by the boundary condition, but U_J^{n+1} must be obtained in some other way; we shall try to deduce an appropriate numerical boundary condition for it from an energy analysis of the scheme's stability.

Suppose we multiply (5.121) by $U_j^{n+1} + U_j^{n-1}$ and sum, to obtain

$$\|U^{n+1}\|^2 - \|U^{n-1}\|^2 = -2\nu\langle U^{n+1} + U^{n-1}, \Delta_{0x}U^n\rangle. \qquad (5.122)$$

As in Example 2 we want to express the right-hand side of (5.122) as the difference between two similar expressions evaluated at successive time levels. To this end we apply the second summation by parts formula (5.104) to obtain, with $V = U^{n-1}$ and $W = U^n$,

$$2\nu\left[\langle U^{n-1}, \Delta_{0x}U^n\rangle + \langle U^n, \Delta_{0x}U^{n-1}\rangle\right]$$
$$= a\Delta t \left[U_{J-1}^{n-1}U_J^n + U_J^{n-1}U_{J-1}^n\right]. \qquad (5.123)$$

The result is

$$\|U^{n+1}\|^2 - \|U^{n-1}\|^2 = -2\nu\langle U^{n+1}, \Delta_{0x}U^n\rangle + 2\nu\langle U^n, \Delta_{0x}U^{n-1}\rangle$$
$$- a\Delta t \left[U_{J-1}^n U_J^n + U_J^{n-1}U_{J-1}^n\right]. \qquad (5.124)$$

The inner product terms are ready for cancellation when we combine successive time steps; but they also involve terms in U_J^n, the value at the right-hand boundary, and we have not yet defined how this is to be calculated. If we collect all the boundary terms involved in (5.124) we obtain

$$-a\Delta t \left[U_{J-1}^{n+1}U_J^n - U_{J-1}^n U_J^{n-1} + U_{J-1}^{n-1}U_J^n + U_J^{n-1}U_{J-1}^n\right].$$

The second and fourth terms cancel here; and then we will obtain a negative definite contribution from the remaining boundary terms if we apply the boundary condition

$$U_J^n = \tfrac{1}{2}(U_{J-1}^{n-1} + U_{J-1}^{n+1}). \qquad (5.125)$$

Thus if we now sum the resulting inequalities over the time levels n, $n-1, \ldots, 2, 1$, and denote by a prime the omission of right-hand boundary terms from the inner products, we obtain

$$\|U^{n+1}\|^2 + \|U^n\|^2 + 2\nu\langle U^{n+1}, \Delta_{0x}U^n\rangle'$$
$$\leq \|U^1\|^2 + \|U^0\|^2 + 2\nu\langle U^1, \Delta_{0x}U^0\rangle'. \qquad (5.126)$$

Finally, we need to apply the Cauchy–Schwarz inequality to these inner products to get

$$2\langle U^{n+1}, \Delta_{0x}U^n\rangle' = \Delta x \sum_{j=1}^{J-1} U_j^{n+1}\left[U_{j+1}^n - U_{j-1}^n\right]'$$
$$\leq \|U^{n+1}\|^2 + \|U^n\|^2, \qquad (5.127)$$

where again we have denoted by a prime the omission of the boundary term from the sum. Both sides of (5.126) are then positive definite, and equivalent to the respective l_2 norms, if we have $\nu < 1$. That is, we have stability in the l_2 norm for such CFL numbers but, as in Example 2, we cannot establish stability in this way right up to $\nu = 1$.

5.10 Summary of the theory

Without doubt, Fourier analysis is the most useful and precise tool for studying the stability, and accuracy, of difference schemes; and, by the Lax Equivalence Theorem, stability is the key property. Yet it is the most restricted tool, for Fourier analysis can be strictly applied only to linear problems, with constant coefficients and periodic boundary conditions, approximated by difference schemes on uniform meshes, and with stability studied only in the l_2 norm. Most of the theoretical developments in the last forty years can therefore be regarded as showing that the conclusions deduced from Fourier analysis are more widely applicable; that this is possible stems from the early observation by von Neumann that most instabilities in difference schemes are initiated by high frequency modes, and are therefore rather local phenomena.

For parabolic problems, we have already seen through a number of examples how a maximum principle can establish stability in the maximum norm for a wider class of problems – with variable coefficients, on non-uniform meshes, and in non-rectangular regions with realistic boundary conditions. Where Fourier analysis can be applied it gives necessary stability conditions, which are sometimes weaker than those given by analysis in the maximum norm. Yet, from the work of Fritz John already referred to, the conditions obtained from the Fourier analysis are sufficient to establish the stability defined in (5.28) which is needed to ensure convergence through Theorem 5.1, and is often called *Lax–Richtmyer stability* because of those authors' seminal survey paper.[1] There are two main reasons for this situation.

The first is that the choice of stability condition in (5.28) is directed towards such a result. In the 1950's there were many alternative definitions of stability in use which could lead to very confusing phenomena; for instance, examples could be given such that, on the one hand, locally derived stability conditions were not sufficient for overall stability, yet other examples showed that such conditions were not necessary. The key

---

[1] Lax, P.D. and Richtmyer, R.D. (1956), Survey of the stability of linear finite difference equations, *Comm. Pure Appl. Math.* **9**, 267–93.

property of (5.28) is that small perturbations to a difference scheme do not change its stability. Based on the studies of Kreiss (1962) already referred to in Section 5.6, and of Strang in 1964,[1] into these stability phenomena, we have the following lemma.

Lemma 5.4 *Suppose the difference scheme* $U^{n+1} = B_1^{-1}B_0U^n$ *is stable, and* $C(\Delta t)$ *is a bounded family of operators. Then the scheme*

$$U^{n+1} = \left[B_1^{-1}B_0 + \Delta t\, C(\Delta t)\right] U^n \tag{5.128}$$

is stable.

Proof Suppose $\|(B_1^{-1}B_0)^n\| \leq K_1$ and $\|C(\Delta t)\| \leq K_2$, and consider the result of multiplying out the product $[B_1^{-1}B_0 + \Delta t\, C]^n$. This will consist of 2^n terms, of which $\binom{n}{j}$ terms involve j factors $\Delta t\, C$ interspersed in $n - j$ factors $B_1^{-1}B_0$; the latter can occur in at most $j + 1$ sequences of consecutive factors, the norm of each sequence being bounded by K_1 and hence the norm of each such term by $K_2^j K_1^{j+1}$. Thus overall we obtain the bound

$$\left\|[B_1^{-1}B_0 + \Delta t\, C]^n\right\| \leq \sum_{j=0}^{n} \binom{n}{j} K_1^{j+1}(\Delta t\, K_2)^j \tag{5.129}$$

$$= K_1(1 + \Delta t\, K_1 K_2)^n \leq K_1 e^{n\Delta t\, K_1 K_2} \tag{5.130}$$

which is bounded for $n\Delta t \leq T$. $\qquad\square$

This can be used for example in the following situation. Suppose $u_t = Lu + a(x)u$, where L is a linear, constant coefficient, operator. Then the stability of the scheme $B_1 U^{n+1} = B_0 U^n$ that approximates $u_t = Lu$ may be analysed by Fourier methods, and the result is unchanged by the addition of the $a(x)u$ term to the problem.

The second factor enabling Fourier analysis to be applied to variable coefficient problems is that many difference schemes are *dissipative*, in the sense that each eigenvalue of the amplification matrix satisfies a relation of the form

$$|\lambda(x, \Delta t, \boldsymbol{\xi})|^2 \leq 1 - \delta|\boldsymbol{\xi}|^{2r}, \tag{5.131}$$

where $\boldsymbol{\xi} = (k_1\Delta x_1,\ k_2\Delta x_2, \ldots, k_d\Delta x_d)^T$, for some $\delta > 0$ and some positive integer r. This was what was exploited by Fritz John, and enabled

---

[1] Strang, G. (1964), Wiener–Hopf difference equations, *J. Math. Mech.* **13**(1), 85–96.

Kreiss in 1964[1] to develop a similar theory for hyperbolic equations. The latter's analysis was in the l_2 norm and made great use of energy methods, so generalising results such as those obtained for very particular cases in Section 5.9.

A more recent influence on the theory has come from developments in the understanding of numerical methods for ODEs, and we might expect this to grow in the future. We have already seen in Theorem 5.3 how a method that is composed of a spatial difference operator incorporated in a Runge–Kutta time-stepping procedure can be shown to be stable if the spatial operator satisfies a coercivity condition. And the examples given in Section 5.9 show the relationship of this result to the energy analysis of typical methods. Another example is provided by the modified equation analysis of Section 5.8. The treatment there shows that this is a very valuable tool for understanding our methods, but at the present time it gives few rigorous results. On the other hand, recent developments of the analysis for ODEs have yielded rigorous error bounds in some cases. An example is provided by the work of Reich[2] where, as is common in that field, it is referred to as backward error analysis. His results exploit the *symplectic* property of a method, which again is an area in ODE theory where there have been very significant developments during the last few years – see the following bibliographic notes for references. In Section 4.9 we have referred briefly to the way in which these ideas have been extended to PDEs to give *multi-symplectic methods*, and we might expect the advantages of such methods to appear through developments in their modified equation analysis.

In a similar way to the understanding of dissipative methods, a combination of energy methods and Fourier analysis has enabled a thorough theory to be developed for the effect of boundary conditions on stability. In Chapter 4 we have already referred to the work of Godunov and Ryabenkii, who gave important necessary stability criteria for discrete boundary conditions, and in Section 4.12 we gave an example of the methods employed. The influential paper[3] by Gustafsson, Kreiss and Sundström established closely related necessary and sufficient conditions for Lax–Richtmyer stability; and many subsequent publications have

---

[1] Kreiss, H.O. (1964), On difference approximations of the dissipative type for hyperbolic differential equations, *Comm. Pure Appl. Math.* **17**, 335–53.

[2] Reich, S.(1999), Backward error analysis for numerical integrators, *SIAM J. Numer. Anal.* **36**, 1549–70.

[3] Gustafsson, B., Kreiss, H.O. and Sundström, A. (1972), Stability theory of difference approximations for mixed initial boundary value problems. II, *Math. Comp.* **26**(119), 649–86.

deduced valuable practical criteria, including those needed for strict or practical stability. One of the important properties of difference schemes that this work highlights, and which we have already referred to in Sections 4.8 and 4.9, is that of *group velocity*. In terms of the amplification factor written as $\lambda = e^{-i\omega\Delta t}$, this is defined as

$$C(k) := \partial\omega/\partial k. \tag{5.132}$$

The reader is referred to the excellent account by Trefethen[1] of its relevance to the stability of boundary conditions.

Finally, a much better understanding of nonlinear problems has been developed in recent years. The starting point is a definition of stability based on the difference between two approximations, as in (5.20). It has long been recognised that the stability of a scheme linearised about the exact solution was necessary for its convergence, but this was quite inadequate to distinguish the very different behaviour of quite similar schemes, when for instance applied to the inviscid Burgers equation $u_t + uu_x = 0$. However, by introducing a concept of *stability thresholds*, López-Marcos and Sanz-Serna[2] and others have been able to develop a result that is very similar to the Lax Equivalence Theorem. Of more practical significance is the concept of *TV-stability* which has developed from the TVD properties of difference schemes described in Section 4.7, and of schemes which satisfy discrete entropy conditions – see LeVeque (1992, 2002).

Bibliographic notes and recommended reading

The standard reference for the material of this chapter remains Richtmyer and Morton (1967), where a Banach space framework is employed to give a complete proof of the Lax Equivalence Theorem. A thorough account of the Kreiss and Buchanan matrix theorems is also given there, together with a number of simpler sufficient stability criteria; but several generalisations of these theorems and much more compact proofs have appeared in the literature in recent years. Moreover, the theory has been developed considerably and the reader is referred to the original papers quoted in Section 5.10 for details of these developments.

A convenient source to consult for surveys of recent developments is the set of the annual volumes of *Acta Numerica*. For example, in the 2003

---

[1] Trefethen, L.N. (1982), Group velocity in finite difference schemes, *SIAM Rev.* **24**, 113–36.

[2] López-Marcos, J.C. and Sanz-Serna, J.M. (1988), Stability and convergence in numerical analysis III: linear investigation of nonlinear stability, *IMA J. Numer. Anal.* **8**(1), 71–84.

volume the articles by Cockburn[1] and Tadmor[2] are both very relevant to the topics discussed in this chapter. Progress with the application of symplectic integration and geometric integration methods, and similarly of modified equation analysis, to PDEs, may also be followed in these volumes.

Exercises

5.1   Consider the DuFort–Frankel scheme for the convection–diffusion equation $u_t + au_x = bu_{xx}$, where $b > 0$, on a uniform mesh,

$$U_j^{n+1} = U_j^{n-1} - 2\nu\Delta_{0x}U_j^n + 2\mu\left(U_{j-1}^n + U_{j+1}^n - U_j^{n+1} - U_j^{n-1}\right)$$

where $\nu = a\Delta t/\Delta x$, $\mu = b\Delta t/(\Delta x)^2$. Show that the scheme is stable for $\nu^2 \leq 1$ with no restriction on the size of μ, but that a restriction on μ is needed to ensure its consistency.

5.2   Consider the problem $u_t = u_{xx}$ on the region $0 < x < 1, t > 0$, with initial condition $u(x,0) = f(x)$ and boundary conditions $u_x(0,t) = 0$ and $u_x(1,t) + u(1,t) = 0$. Show that when the derivative boundary conditions are approximated by central differences the explicit method leads to the system

$$U_j^{n+1} = (1 - 2\mu)U_j^n + \mu(U_{j-1}^n + U_{j+1}^n), \quad j = 1, 2, \ldots, J - 1,$$
$$U_0^{n+1} = (1 - 2\mu)U_0^n + 2\mu U_1^n,$$
$$U_J^{n+1} = (1 - 2\mu - 2\mu\Delta x)U_J^n + 2\mu U_{J-1}^n.$$

Show that $V_j^n = \lambda^n \cos kj\Delta x$ satisfies these equations provided that

$$\lambda = 1 - 4\mu\sin^2\tfrac{1}{2}k\Delta x,$$

and that k satisfies

$$\Delta x \cot k = \sin k\Delta x.$$

By drawing graphs of $\cot k$ and $\sin k\Delta x$ show that there are only J real solutions of this equation which lead to different mesh functions V_j^n. Verify that there is also one complex solution which is approximately $k\Delta x = \pi + iy\Delta x$ when Δx is small,

[1] Cockburn, B. (2003), Continuous dependence and error estimation for viscosity methods, *Acta Numerica.* **12**, 127–80.

[2] Tadmor, E. (2003), Entropy stability theory for difference approximations of nonlinear conservation laws and related time-dependent problems, *Acta Numerica* **12**, Cambridge University Press, 451–512.

where y is the unique positive root of $y = \coth y$. Show that, with this value of k, $\lambda = 1 - 4\mu \cosh^2 \frac{1}{2} y \Delta x$. Deduce that for this problem the explicit method with $\mu = \frac{1}{2}$ gives an error growth, while still being stable by the criteria of (2.55) and (2.56).

5.3　Consider the quadratic polynomial $p(z) = az^2 + 2bz + c$, where a, b, and c are complex.

(i) Show that if $|c| > |a|$ then at least one of the zeros of $p(z)$ has modulus greater than 1.

(ii) Suppose that $|c| < |a|$, and define $q(z) = \bar{c}z^2 + 2\bar{b}z + \bar{a}$ and $r(z) = \bar{a}p(z) - cq(z)$. Show that if $|z| = 1$ then $|q(z)| = |p(z)|$ and hence that $|cq(z)| < |\bar{a}p(z)|$. Deduce from Rouché's theorem that $p(z)$ and $r(z)$ have the same number of zeros inside the unit circle, and hence that both zeros of $p(z)$ have modulus less than or equal to 1 if and only if $2|\bar{a}b - \bar{b}c| \le |a|^2 - |c|^2$,

(iii) Now suppose that $|a| = |c|$. Write $c = ae^{i\theta}$, $b = a\beta e^{i\phi}$, $z = ue^{i\theta/2}$, where β is real, and show that

$$u^2 + 2\beta e^{i(\phi - \theta/2)} u + 1 = 0.$$

Show that the roots of this equation lie on the unit circle if and only if $\theta = 2\phi$ and $|\beta| \le 1$. Deduce that if $|c| = |a|$ the zeros of $p(z)$ have modulus less than or equal to 1 if and only if $\bar{a}b = \bar{b}c$ and $|b| \le |a|$.

5.4　The convection–diffusion problem $u_t + au_x = bu_{xx}$ is approximated on the whole real line by the MacCormack scheme

$$U^{n+*} = U^n - \nu \Delta_- U^n,$$
$$U^{n+1} = U^n - \tfrac{1}{2}\nu(\Delta_- U^n + \Delta_+ U^{n+*}) + \tfrac{1}{2}\mu(\delta^2 U^n + \delta^2 U^{n+*}),$$

where $\nu = a\Delta t / \Delta x$ and $\mu = b\Delta t / (\Delta x)^2$ are positive. Find the conditions under which the von Neumann stability condition is satisfied. Show that for practical stability it is necessary that

$$2\mu - 1 \le \nu \le 1,$$

and that when $\mu = \frac{1}{2}$ this is also a sufficient condition.

5.5　Find the necessary and sufficient stability criterion for

$$U^{n+1} - U^n = -\mu \delta^2 V^n,$$
$$V^{n+1} - V^n = \mu \delta^2 U^{n+1}, \qquad \mu = a\Delta t / (\Delta x)^2,$$

which represents the vibration of a bar.

Suppose now the bar is subject to tension and terms $\nu\Delta_0 V^n$ and $\nu\Delta_0 U^{n+1}$, with $\nu = b\Delta t/\Delta x$, are introduced into the two equations. Is Lax–Richtmyer stability affected? Find a simple sufficient criterion for practical stability.

5.6   Suppose that the Lax–Wendroff method is applied on a uniform mesh to the advection equation $u_t + au_x = 0$, with constant a. Show that, over the whole real line, with $\nu = a\Delta t/\Delta x$,

$$||U^{n+1}||^2 = ||U^n||^2 - \tfrac{1}{2}\nu^2(1-\nu^2)\left[||\Delta_- U^n||^2 - \langle \Delta_- U^n, \Delta_+ U^n\rangle\right]$$

and hence deduce the stability condition.

Now If the method is applied on the interval $(0,1)$ with $a > 0$ and the boundary condition $U_0^n = 0$ at $x = 0$, find a simple boundary condition at $x = 1$ which will be stable.

5.7   Suppose the box scheme is used for the solution of $u_t + au_x = 0$, where a is a positive constant, in the region $x > 0$, $t > 0$ with a uniform mesh. The boundary conditions give $u(x,0) = f(x)$ for $x > 0$, where $f \in L_2(0,\infty)$, and $u(0,t) = 0$ for $t > 0$. Writing the scheme in the form

$$U_j^{n+1} + U_{j-1}^{n+1} + \nu(U_j^{n+1} - U_{j-1}^{n+1}) = U_j^n + U_{j-1}^n - \nu(U_j^n - U_{j-1}^n),$$

and defining

$$S_n := \sum_{j=1}^{\infty}\left\{(U_j^n + U_{j-1}^n)^2 + \nu^2(U_j^n - U_{j-1}^n)^2\right\},$$

where $\nu = a\Delta t/\Delta x$, show that $S_{n+1} = S_n$. Deduce that the numerical solution is bounded in $l_2(0,\infty)$ for all Δx and Δt.

Now suppose that a is a positive function of x with $|a'(x)| \leq L$, and that in the box scheme a is replaced by $a(x_j - \tfrac{1}{2}\Delta x)$. Show that

$$S_{n+1} - S_n \leq 2L\Delta t\left\{\sum_{j=1}^{\infty}[(U_j^n)^2 + (U_j^{n+1})^2]\right\}.$$

What does this imply about the stability of the scheme?

5.8   Suppose the θ-method is applied to $u_t = u_{xx}$ on a uniform mesh with the boundary conditions $U_J^n = 0$ on the right and $aU_0^n + bU_1^n = 0$ on the left, for some constants a and b. Consider solution modes of the form $U_j^n = \widehat{U}\lambda^n\mu^j$ to the difference equations at interior points and find the relation needed between λ and μ.

Show that there can be no solution with $|\lambda| > 1$ and $|\mu| < 1$ that satisfies the boundary condition on the left, either if $|a/b| \geq 1$ or if $|a/b| < 1$ and

$$\frac{2\Delta t(1 - 2\theta)}{(\Delta x)^2} \leq \frac{4ab}{(a + b)^2}.$$

[NB. The Godunov–Ryabenkii necessary conditions for stability require that any solution mode of the form $\lambda^n \mu^j$ to the interior equations satisfies: (i) if $|\mu| = 1$ then $|\lambda| \leq 1$; (ii) if $|\mu| < 1$ and $|\lambda| > 1$ then boundary conditions are applied on the left to remove the mode; (iii) similarly on the right for $|\mu| > 1$.]

6

Linear second order elliptic equations in two dimensions

6.1 A model problem

As in previous chapters we shall begin with the simplest model problem, to solve

$$u_{xx} + u_{yy} + f(x, y) = 0, \quad (x, y) \in \Omega, \tag{6.1a}$$

$$u = 0, \quad (x, y) \in \partial\Omega, \tag{6.1b}$$

where Ω is the unit square

$$\Omega := (0, 1) \times (0, 1) \tag{6.2}$$

and $\partial\Omega$ is the boundary of the square. If we compare with the parabolic equation, of the type discussed in Chapter 3,

$$\frac{\partial u}{\partial t} = u_{xx} + u_{yy} + f(x, y) \tag{6.3}$$

we see that, if the solution converges to a limit as $t \to \infty$, this limit will be the solution of (6.1a). This connection between the elliptic problem and the time-dependent solution of the parabolic problem has often been exploited in the solution of elliptic problems; in Chapter 7 we shall discuss the relation between iterative methods for the solution of elliptic problems and time stepping finite difference methods for the solution of the corresponding parabolic problems.

We cover the unit square with a uniform square grid with J intervals in each direction, so that

$$\Delta x = \Delta y = 1/J, \tag{6.4}$$

and we approximate (6.1) by the central difference scheme

$$\frac{U_{r+1,s} + U_{r-1,s} + U_{r,s+1} + U_{r,s-1} - 4U_{r,s}}{(\Delta x)^2} + f_{r,s} = 0. \tag{6.5}$$

Writing equation (6.5) with $r = 1, 2, \ldots, J-1$ and $s = 1, 2, \ldots, J-1$ we obtain a system of $(J-1)^2$ equations which have exactly the same structure as the systems which arose in the solution of parabolic equations by implicit methods in Chapter 3. We shall assume for the moment that the equations have been solved in some way, and go on to investigate the accuracy of the result.

6.2 Error analysis of the model problem

As usual we begin with the truncation error. Substituting into equation (6.5) the exact solution $u(x_r, y_s)$ of the differential equation (6.1), and expanding in Taylor series, the truncation error is easily found to be

$$T_{r,s} = \tfrac{1}{12}(\Delta x)^2 \left(u_{xxxx} + u_{yyyy}\right)_{r,s} + o\left((\Delta x)^2\right). \tag{6.6}$$

Indeed this is bounded by T, where

$$|T_{r,s}| \leq T := \tfrac{1}{12}(\Delta x)^2 \left(M_{xxxx} + M_{yyyy}\right) \tag{6.7}$$

in the usual notation for bounds on the partial derivatives of $u(x, y)$.

We now define an operator L_h on the set of all arrays U of values $U_{r,s}$,

$$(L_hU)_{r,s} \equiv L_hU_{r,s}$$
$$:= \frac{1}{(\Delta x)^2}\left(U_{r+1,s} + U_{r-1,s} + U_{r,s+1} + U_{r,s-1} - 4U_{r,s}\right), \tag{6.8}$$

at all the interior points $J_\Omega \equiv \{(x_r, y_s); r = 1, 2, \ldots, J-1 \text{ and } s = 1, 2, \ldots, J-1\}$. Then the numerical approximation satisfies the equation

$$L_hU_{r,s} + f_{r,s} = 0, \tag{6.9}$$

and the exact solution satisfies

$$L_hu_{r,s} + f_{r,s} = T_{r,s}. \tag{6.10}$$

We define the error in the usual way as

$$e_{r,s} := U_{r,s} - u_{r,s} \tag{6.11}$$

and see that

$$L_he_{r,s} = -T_{r,s}. \tag{6.12}$$

Since the values of $u(x, y)$ were given at all points on the boundary, it follows that the boundary values of $e_{r,s}$ are zero.

To obtain a bound on $e_{r,s}$ we first define a *comparison function*

$$\Phi_{r,s} := (x_r - \tfrac{1}{2})^2 + (y_s - \tfrac{1}{2})^2. \tag{6.13}$$

Then

$$L_h \Phi_{r,s} = 4; \tag{6.14}$$

this can be shown either by direct calculation, or more easily by noting that Φ is a quadratic function of x and y, and therefore $L_h \Phi$ must give the exact value of $\Phi_{xx} + \Phi_{yy}$, which is evidently 4. If we now write

$$\psi_{r,s} := e_{r,s} + \tfrac{1}{4} T \Phi_{r,s} \tag{6.15}$$

we obtain

$$\begin{aligned}
L_h \psi_{r,s} &= L_h e_{r,s} + \tfrac{1}{4} T L_h \Phi_{r,s} \\
&= -T_{r,s} + T \\
&\geq 0 \quad \forall (x_r, y_s) \in J_\Omega.
\end{aligned} \tag{6.16}$$

We now appeal to a maximum principle, similar to that used in Theorem 2.2 in Chapter 2, and which we shall prove in a more general context in Section 6.5. Briefly, the operator L_h is such that, if at a point (x_r, y_s) we have $L_h \psi_{r,s} \geq 0$, then $\psi_{r,s}$ cannot be greater than all the neighbouring values. Hence it follows from (6.16) that a positive maximum value of ψ must be attained at a point on the boundary of our square region. But $e_{r,s}$ vanishes on the boundary, and the maximum of Φ is $\tfrac{1}{2}$, being attained at the corners. Hence the maximum of ψ on the boundary is $\tfrac{1}{8} T$, and

$$\psi_{r,s} \leq \tfrac{1}{8} T \quad \forall (x_r, y_s) \in J_\Omega. \tag{6.17}$$

But since Φ is nonnegative, from the definition of ψ we obtain

$$\begin{aligned}
U_{r,s} - u(x_r, y_s) = e_{r,s} &\leq \psi_{r,s} \\
&\leq \tfrac{1}{8} T \\
&= \tfrac{1}{96} (\Delta x)^2 \left(M_{xxxx} + M_{yyyy} \right).
\end{aligned} \tag{6.18}$$

Notice that this is a one-sided bound; it is a simple matter to repeat the analysis, but defining $\psi_{r,s} = \tfrac{1}{4} T \Phi_{r,s} - e_{r,s}$. The result will be to show that $-e_{r,s} \leq \tfrac{1}{8} T$, from which we finally obtain the required bound

$$|U_{r,s} - u(x_r, y_s)| \leq \tfrac{1}{96} (\Delta x)^2 (M_{xxxx} + M_{yyyy}). \tag{6.19}$$

6.3 The general diffusion equation

We shall now extend this approach to a more general elliptic problem, the diffusion equation

$$\nabla \cdot (a\nabla u) + f = 0 \quad \text{in } \Omega, \tag{6.20}$$

where

$$a(x, y) \geq a_0 > 0. \tag{6.21}$$

We shall suppose that Ω is a bounded open region with boundary $\partial\Omega$. The boundary conditions may have the general form

$$\alpha_0 u + \alpha_1 \partial u / \partial n = g \quad \text{on } \partial\Omega, \tag{6.22}$$

where $\partial/\partial n$ represents the derivative in the direction of the outward normal and

$$\alpha_0 \geq 0, \quad \alpha_1 \geq 0, \quad \alpha_0 + \alpha_1 > 0. \tag{6.23}$$

As in the previous chapter we cover the region Ω with a regular mesh, with size Δx in the x-direction and Δy in the y-direction.

Suppose a is smoothly varying and write $b = \partial a/\partial x$, $c = \partial a/\partial y$. Then we could expand (6.20) as

$$a\nabla^2 u + bu_x + cu_y + f = 0. \tag{6.24}$$

At points away from the boundary we can approximate this equation by using central differences, giving an approximation $U := \{U_{r,s}, (r, s) \in J_\Omega\}$ satisfying

$$a_{r,s}\left[\frac{\delta_x^2 U_{r,s}}{(\Delta x)^2} + \frac{\delta_y^2 U_{r,s}}{(\Delta y)^2}\right] + b_{r,s}\left[\frac{\Delta_{0x} U_{r,s}}{\Delta x}\right] + c_{r,s}\left[\frac{\Delta_{0y} U_{r,s}}{\Delta y}\right] + f_{r,s} = 0. \tag{6.25}$$

The truncation error of this five-point scheme is defined in the usual way and is easily found to be second order in $\Delta x, \Delta y$. The terms in this equation which involve $U_{r+1,s}$ and $U_{r-1,s}$ are

$$\left[\frac{a_{r,s}}{(\Delta x)^2} - \frac{b_{r,s}}{2\Delta x}\right] U_{r-1,s} + \left[\frac{a_{r,s}}{(\Delta x)^2} + \frac{b_{r,s}}{2\Delta x}\right] U_{r+1,s}. \tag{6.26}$$

In order to use a maximum principle to analyse the error of this scheme it is necessary to ensure that all the coefficients at the points which are neighbours to (r, s) have the same sign. This would evidently require that

$$|b_{r,s}|\Delta x \leq 2a_{r,s} \qquad \forall r, s \qquad (6.27)$$

with a similar restriction for $c_{r,s}$. This would imply the use of a fine mesh where the diffusion coefficient $a(x, y)$ is small but changing rapidly.

A more natural scheme however is based more directly on an integral form of (6.20), as we saw when considering polar co-ordinates in Section 2.8. Consider the control volume V around a mesh point, which was introduced in the last chapter and is indicated by dotted lines in Fig. 6.1. Integrating (6.20) over this volume and using Gauss' theorem we obtain

$$\int_{\partial V} a(\partial u/\partial n)\, \mathrm{d}l + \int_V f\, \mathrm{d}x\, \mathrm{d}y = 0. \qquad (6.28)$$

We can now construct a difference scheme by approximating the terms in (6.28). The boundary ∂V of V has normals in the co-ordinate directions and the normal derivatives can be approximated by divided differences using the same five points as in (6.25). We approximate each of the line integrals along the four sides of the rectangle by the product of the length of the side and the value of the normal derivative at the mid-point of the side. In the same way, we approximate the integral of $f(x, y)$ over the element by the product of the area of the element and the value of $f(x, y)$ at the centre. As a result, we obtain the scheme

$$\frac{\Delta y}{\Delta x} \left[a_{r+1/2,s} \left(U_{r+1,s} - U_{r,s} \right) - a_{r-1/2,s} \left(U_{r,s} - U_{r-1,s} \right) \right]$$

$$+ \frac{\Delta x}{\Delta y} \left[a_{r,s+1/2} \left(U_{r,s+1} - U_{r,s} \right) - a_{r,s-1/2} \left(U_{r,s} - U_{r,s-1} \right) \right]$$

$$+ \Delta x \Delta y f_{r,s} = 0 \qquad (6.29a)$$

or

$$\left[\frac{\delta_x(a\delta_x U)}{(\Delta x)^2} + \frac{\delta_y(a\delta_y U)}{(\Delta y)^2} \right]_{r,s} + f_{r,s} = 0. \qquad (6.29b)$$

It is often convenient to use the 'compass point' notation indicated in Fig. 6.1 and to write (6.29a), (6.29b) as

$$\frac{a_e(U_E - U_P) - a_w(U_P - U_W)}{(\Delta x)^2} + \frac{a_n(U_N - U_P) - a_s(U_P - U_S)}{(\Delta y)^2} + f_P = 0. \qquad (6.30)$$

Since we have assumed that the function $a(x, y)$ is positive it is now easy to see that the coefficients in the scheme (6.30) always have the correct sign, without any restriction on the size of the mesh.

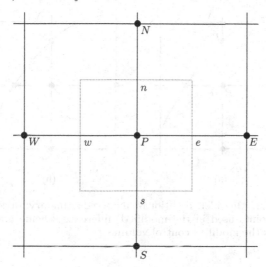

Fig. 6.1. The control volume about the point P.

Another advantage of the form (6.29), comes in problems where there is an interface at which material properties (represented by a) change abruptly, yet the normal flux $a\partial u/\partial n$ is continuous. Suppose we arrange the grid so that the material interface passes vertically through the point e, with the constant value a_E holding on the right and the value a_P on the left. Then in terms of an intermediate value U_e on the interface, the common value of the flux can be approximated by

$$\frac{a_E(U_E - U_e)}{\frac{1}{2}\Delta x} = \frac{a_P(U_e - U_P)}{\frac{1}{2}\Delta x}$$
$$= \frac{a_e(U_E - U_P)}{\Delta x}, \text{ say,} \qquad (6.31)$$

where the elimination of U_e to give the last expression, ready for substitution into (6.30), is accomplished by defining a_e through

$$\frac{2}{a_e} = \frac{1}{a_E} + \frac{1}{a_P}. \qquad (6.32)$$

6.4 Boundary conditions on a curved boundary

Either form (6.25) or (6.30) has to be modified in the neighbourhood of a boundary which does not lie along one of the mesh lines. We saw in Section 3.4 how the second derivatives can be approximated at such

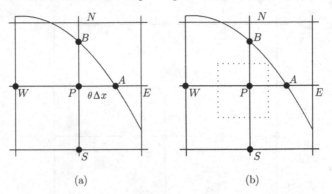

Fig. 6.2. Dirichlet condition on a curved boundary; (a) shows the points used in the modified difference scheme and (b) shows the modified control volume.

points, and now have to extend the method to the more general diffusion equation.

Let us first consider a situation such as that in Fig. 6.2 where Dirichlet data are given on the curved boundary. If we use Taylor series expansions to obtain an approximation to (6.24) at the point P in the form (6.25), we need to use the given values at A and B instead of the values at E and N. It is sufficient to consider A only, and we write $PA = \theta \Delta x$. Then Taylor expansions for u_A and u_W give

$$u_A = \left[u + \theta \Delta x\, u_x + \tfrac{1}{2}(\theta \Delta x)^2 u_{xx} + \cdots \right]_P,$$
$$u_W = \left[u - \Delta x\, u_x + \tfrac{1}{2}(\Delta x)^2 u_{xx} - \cdots \right]_P.$$

From these, by eliminating first u_{xx} and then u_x, we obtain approximations

$$[u_x]_P \approx \frac{u_A - \theta^2 u_W - (1 - \theta^2) u_P}{\theta(1 + \theta)\Delta x}, \tag{6.33a}$$

$$[u_{xx}]_P \approx \frac{u_A + \theta u_W - (1 + \theta) u_P}{\tfrac{1}{2}\theta(1 + \theta)(\Delta x)^2}. \tag{6.33b}$$

Carrying out the same construction for U_P, U_S and U_B, we readily obtain an appropriately modified form of (6.25). For a maximum principle to hold we shall again require a restriction on the size of the mesh, just as at an ordinary interior point.

The alternative integral form of the difference scheme (6.30) is modified near a Dirichlet boundary in a similar way. Thus for each mesh point we draw a rectangular control volume about it as in Fig. 6.1 but, if the mesh line joining P to one of its neighbours crosses the boundary, the corresponding side of the rectangle is drawn perpendicular to the mesh line, crossing it half-way to the boundary, as in Fig. 6.2(b). For example, the distance PA in this figure is a fraction θ of the mesh size $PE = \Delta x$, so the width of the volume V is $\frac{1}{2}(1+\theta)\Delta x$.

Hence the line integral along the bottom side of the element is approximated by

$$\frac{1}{2}(1+\theta)\Delta x \left[\frac{-a_s(U_P - U_S)}{\Delta y} \right]. \qquad (6.34)$$

We must also make an adjustment to the approximation to the normal derivative in the line integral up the right-hand side of the element; this derivative is approximated by

$$\frac{a_a(U_A - U_P)}{\theta \Delta x}, \qquad (6.35)$$

where a_a is the value of $a(x,y)$ at the point midway between A and P and a_b will have a similar meaning. Noting that in Fig. 6.2(b) the boundary cuts the mesh lines at two points, A and B with $PA = \theta \Delta x$ and $PB = \phi \Delta y$, we obtain the difference approximation at the point P as

$$\frac{1}{2}(1+\phi)\Delta y \left[\frac{a_a(U_A - U_P)}{\theta \Delta x} - \frac{a_w(U_P - U_W)}{\Delta x} \right]$$
$$+ \frac{1}{2}(1+\theta)\Delta x \left[\frac{a_b(U_B - U_P)}{\phi \Delta y} - \frac{a_s(U_P - U_S)}{\Delta y} \right]$$
$$+ \frac{1}{4}(1+\theta)(1+\phi)\Delta x \Delta y f_P = 0. \qquad (6.36)$$

It is clear that in the case where $a(x,y)$ is constant this scheme will be identical to that given by (6.33a, b). More generally, it has the advantage that the coefficients still satisfy the conditions required for a maximum principle.

Derivative boundary conditions are more difficult to deal with. As we saw in Chapter 3, it is difficult to construct accurate difference approximations to the normal derivative, and it is necessary to take account of a number of different possible geometrical configurations. Moreover, we shall see in Section 6.7 that derivative boundary conditions are dealt with much more straightforwardly in a finite element approximation. However we will show how in a simple case the integral form of the equation can be adapted.

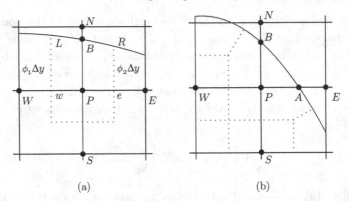

(a)   (b)

Fig. 6.3. Integral form of Neumann condition; the boundary
intersects one mesh line in (a) and two in (b).

Consider the situation shown in Fig. 6.3(a) where the boundary cuts
the mesh line PN at the point B, but does not cut the other lines joining
P to its nearest neighbours. We also assume that a Neumann condition,

$$\frac{\partial u}{\partial n} = g, \qquad (6.37)$$

is given on this part of the boundary; the general condition (6.22) is more
difficult to deal with. We construct an element V round P by drawing
three sides of the rectangle as we have done before, but extending the
two vertical sides to meet the boundary at the points L and R and using
the part of the boundary between them. Writing

$$wL = \phi_1 \Delta y, \quad eR = \phi_2 \Delta y, \qquad (6.38)$$

we can approximate the line integrals along the horizontal and vertical
sides of the element just as we did before, but we notice that the results
will be less accurate, since we are not evaluating the normal derivatives
at the mid-points of the vertical sides. In the same way the double
integral is approximated by the product of the area of the element and
the value of f at P. Again there is a loss of accuracy, since P is not at
the centre of the element.

Finally we must approximate the line integral along the boundary
RBL; here we can write

$$\int a \frac{\partial u}{\partial n} \mathrm{d}l = \int a \, g \, \mathrm{d}l \qquad (6.39)$$

which is approximated by

$$a_B \, g_B \, \psi \Delta x \tag{6.40}$$

where $\psi \Delta x$ is the length of the line LR. This leads finally to the difference equation

$$(\phi_2 + \tfrac{1}{2})\Delta y \left[a_e \frac{U_E - U_P}{\Delta x} \right] + (\phi_1 + \tfrac{1}{2})\Delta y \left[a_w \frac{U_W - U_P}{\Delta x} \right]$$

$$+ \Delta x \left[a_s \frac{U_S - U_P}{\Delta y} \right] + \psi \Delta x \, a_B g_B + \tfrac{1}{2}(\phi_1 + \phi_2 + 1)\Delta x \, \Delta y \, f_P = 0. \tag{6.41}$$

In the fairly common situation where the boundary cuts two of the mesh lines from P to its neighbours, the procedure is much the same, but it will be necessary to use some diagonal lines in the construction of the element, as in Fig. 6.3(b). The geometrical details are tedious, and we shall not discuss them further.

6.5 Error analysis using a maximum principle

We shall assume that in approximating (6.20), (6.22), or any linear elliptic equation, we have been successful in constructing, at each interior point $P \in J_\Omega$, an approximation of the form

$$L_h U_P + f_P + g_P = 0, \tag{6.42}$$

where g_P represents any boundary data that are picked up from other than Dirichlet boundary conditions. We assume that the following conditions hold:–

(i) For each $P \in J_\Omega$, L_h has the form

$$L_h U_P = \sum_k c_k U_k - c_P U_P, \tag{6.43}$$

where the coefficients are positive and the sum over k is taken over mesh points which are neighbours of P. In the difference scheme above these have only included the four nearest neighbours, but the analysis can be applied to more general schemes which involve neighbours such as U_{NE} in a diagonal direction. Also, when P is near the boundary, some of the neighbours may be in a set $J_{\partial\Omega}$ of points on the boundary such as A and B in Fig. 6.2, and equation (6.36), and then the corresponding values U_k are given by Dirichlet boundary conditions. The key requirement is that all the coefficients which occur in (6.43) must be positive.

(ii) For each $P \in J_\Omega$

$$c_P \geq \sum_k c_k. \tag{6.44}$$

(iii) The set J_Ω is *connected*. We say that a point P is *connected* to each of its neighbours that occurs in (6.43) with a non-zero coefficient. The set is then connected if, given any two points P and Q in Ω_J, there is a sequence of points $P = P_0, P_1, P_2, \ldots, P_m = Q$ such that each point P_r is connected to P_{r-1} and P_{r+1}, for $r = 1, 2, \ldots, m - 1$.

(iv) At least one of the equations (6.43) must involve a boundary value U_A which is given by a Dirichlet boundary condition. In other words, Dirichlet boundary conditions must be given on at least part of the boundary. (Such a condition is needed, or at least (6.22) with $\alpha_0 \neq 0$, to ensure uniqueness of the solution of (6.1).)

Note that in these conditions we refer to *interior* points J_Ω and *boundary* points $J_{\partial\Omega}$ in a specific sense. An interior point is one at which an equation of the form (6.42) is satisfied, with the coefficients satisfying the appropriate conditions. A boundary point is a point which appears in at least one equation but is one at which no equation is given and the value of U is prescribed, by a Dirichlet condition. At a point on the boundary of the region at which a Neumann or mixed condition is given we will normally eliminate the value of U, as in the situation illustrated in Fig. 6.3 where the unknown value U_B does not appear in the equation (6.41) for U_P, or in any of the equations; or we treat it as an interior point with its own equation, as in the treatment of the symmetry boundary condition in the example of Section 6.9 below. The *boundary* points are therefore those at which values are prescribed by Dirichlet conditions. Note that these definitions and conventions, which are convenient for our present purposes, are slightly different from those used in Chapter 5 (where difference operators as in (5.8) involved only points in J_Ω) and from those to be used in Chapter 7.

Lemma 6.1 (Maximum Principle) *Suppose that L_h, J_Ω and $J_{\partial\Omega}$ satisfy all the above assumptions and that a mesh function U satisfies*

$$L_h U_P \geq 0 \quad \forall \, P \in J_\Omega. \tag{6.45}$$

Then U cannot attain a nonnegative maximum at an interior point, i.e.

$$\max_{P \in J_\Omega} U_P \leq \max\{ \max_{A \in J_{\partial\Omega}} U_A, 0\}. \tag{6.46}$$

Proof We prove this by contradiction. Suppose the interior maximum $M_\Omega \geq 0$ occurs at a point P and that $M_\Omega > M_{\partial\Omega}$, the maximum over the boundary points. Then from (6.45), (6.43) and (6.44)

$$M_\Omega = U_P < \frac{1}{c_P} \sum_k c_k U_k$$

$$\leq \frac{1}{c_P} \sum_k c_k M_\Omega \leq M_\Omega. \tag{6.47}$$

This means that equality holds throughout (6.47), which implies that all the values U_k involved are also equal to M_Ω. Hence the maximum value is attained also at all the points which are connected neighbours of P. The same argument can then be used at each of these points, and so on. Since we have assumed that J_Ω is connected this shows that U takes the same value M_Ω at all the interior points, and at least one of these points must have a connected neighbour which is on the boundary. This contradicts the assumption that $M_\Omega > M_{\partial\Omega}$. $\qquad\square$

Corollary *The inequality*

$$L_h U_P \leq 0 \quad \forall P \in J_\Omega$$

implies

$$\min_{P \in J_\Omega} U_P \geq \min\{ \min_{A \in J_{\partial\Omega}} U_A, 0\}. \tag{6.48}$$

The proof follows in the same manner as above, or alternatively we may simply apply the lemma to the mesh function $-U$.

We next define the truncation error at each interior point in the usual way, noting that (6.42) has been properly scaled,

$$T_P := L_h u_P + f_P + g_P. \tag{6.49}$$

Hence for the error $e_P = U_P - u_P$ at each interior point we have

$$L_h e_P = -T_P, \quad P \in J_\Omega. \tag{6.50}$$

We shall assume as usual that $e_A = 0$ at all boundary points at which a Dirichlet condition is prescribed, i.e. Dirichlet conditions are imposed exactly. This will enable us to bound e_P by means of the maximum principle in Lemma 6.1.

Theorem 6.1 *Suppose a nonnegative mesh function Φ is defined on $J_\Omega \cup J_{\partial\Omega}$ such that*

$$L_h \Phi_P \geq 1 \quad \forall P \in J_\Omega, \tag{6.51}$$

and that all the above four conditions are satisfied. Then the error in the approximation (6.42) is bounded by

$$|e_P| \leq \left[\max_{A \in J_{\partial\Omega}} \Phi_A \right] \left[\max_{P \in J_\Omega} |T_P| \right]. \tag{6.52}$$

Proof Let us denote by T the maximum of the absolute value of the truncation error, $|T_P|$. Then we can apply Lemma 6.1 to the function $T\Phi_P + e_P$ because

$$L_h(T\Phi_P + e_P) \geq T - T_P \geq 0. \tag{6.53}$$

We therefore deduce, using the fact that Φ is nonnegative, that

$$\max_{P \in J_\Omega}(e_P) \leq \max_{P \in J_\Omega}(T\Phi_P + e_P) \leq \max_{A \in J_{\partial\Omega}}(T\Phi_A + e_A),$$

i.e.,

$$\max_{P \in J_\Omega} e_P \leq \left[\max_{A \in J_{\partial\Omega}} \Phi_A \right] T, \tag{6.54}$$

where we have used the fact that $e_A = 0$ at all boundary points, because of the assumption that we use the exact values of the given Dirichlet boundary conditions.

This gives a one-sided bound. If we apply the lemma to the function $T\Phi_P - e_P$ we obtain a similar bound for $-e_P$, thus giving the required result. $\square$

The presentation and analysis here have been given in rather general terms. The analysis of the model problem in Section 6.2 followed the same method, the only difference being that we used a comparison function Φ for which $L_h\Phi_P = 4$; a more general theorem that includes both cases will be given below. When applying the technique to a particular problem we must find a bound on the truncation error, giving T, and then construct a function Φ. Evidently the determination of T is quite straightforward, requiring simply a Taylor series expansion, whereas it may be more difficult to construct a suitable Φ. This function is not, of course, unique. In the model problem we could, for example, have defined

$$\Phi_{r,s} = \tfrac{1}{4}\left[(x_r - p)^2 + (y_s - q)^2 \right]; \tag{6.55}$$

the required conditions would be satisfied for any values of the constants p and q. The particular values that were chosen, $p = q = \tfrac{1}{2}$, give the smallest value for $\max(\Phi_A)$.

We shall now apply this general analysis to some more complicated cases. Consider first the solution of Poisson's equation (6.1) in a region with a curved boundary, as discussed in Section 6.4. The finite difference approximation will satisfy a maximum principle, and for all points at which all the four neighbours are also interior mesh points, the truncation error has the form (6.6) and satisfies (6.7). However, at mesh points next to the boundary, where one or more of the neighbours lie on the boundary, we must use a more general difference approximation, such as (6.33b). A Taylor series expansion easily shows that

$$\frac{u_A + \theta u_W - (1+\theta)u_P}{\frac{1}{2}\theta(1+\theta)(\Delta x)^2} = (u_{xx})_P - \tfrac{1}{3}(1-\theta)\Delta x\,(u_{xxx})_P + O\left((\Delta x)^2\right).$$
(6.56)

It is possible for an interior mesh point to have more than one neighbour on the boundary, but since $0 < \theta < 1$ in (6.56) it is easy to see that in all cases we can choose positive constants K_1 and K_2 such that

$$|T_{r,s}| \le K_1(\Delta x)^2 \quad \text{at ordinary points,} \tag{6.57a}$$
$$|T_{r,s}| \le K_2\Delta x \quad \text{next to the boundary,} \tag{6.57b}$$

provided that Δx is sufficiently small. Hence

$$|T_{r,s}| \le K_1(\Delta x)^2 + K_2\Delta x \tag{6.58}$$

at all interior mesh points.

Now suppose that the region is contained in a circle with centre (p,q) and radius R, and define the comparison function $\Phi_{r,s}$ by (6.55). Then $L_h\Phi_P = 1$ at all ordinary interior mesh points, as before; this result also holds at points next to the boundary, since the truncation error in (6.56) involves the third and higher derivatives, and vanishes for a quadratic polynomial. We can therefore apply Theorem 6.1 and deduce that

$$|U_{r,s} - u(x_r, y_s)| \le \tfrac{1}{4}R^2[K_1(\Delta x)^2 + K_2\Delta x], \tag{6.59}$$

since $0 \le \Phi \le \tfrac{1}{4}R^2$ throughout the region, and on the boundary. This shows that the error is $O(\Delta x)$ as the mesh size tends to zero, rather than the $O\left((\Delta x)^2\right)$ obtained for a simple region.

A slightly modified comparison function can however be used to produce a sharper error bound. In analysis of problems of this kind it is quite common for the truncation error to have one form at the ordinary points of the mesh, but to be different near the boundary. It is therefore convenient to have a generalised form of Theorem 6.1.

Theorem 6.2 *Suppose that, in the notation of Theorem 6.1, the set J_Ω is partitioned into two disjoint sets*

$$J_\Omega = J_1 \cup J_2, \quad J_1 \cap J_2 = \emptyset;$$

the nonnegative mesh function Φ is defined on $J_\Omega \cup J_{\partial\Omega}$ and satisfies

$$
\begin{aligned}
L_h \Phi_P &\geq C_1 > 0 & \forall\, P \in J_1, \\
L_h \Phi_P &\geq C_2 > 0 & \forall\, P \in J_2;
\end{aligned}
\tag{6.60}
$$

and the truncation error of the approximation (6.42) satisfies

$$
\begin{aligned}
|T_P| &\leq T_1 & \forall\, P \in J_1, \\
|T_P| &\leq T_2 & \forall\, P \in J_2.
\end{aligned}
\tag{6.61}
$$

Then the error in the approximation is bounded by

$$
|e_P| \leq \left[\max_{A \in J_{\partial\Omega}} \Phi_A \right] \max\left\{ \frac{T_1}{C_1}, \frac{T_2}{C_2} \right\}.
\tag{6.62}
$$

Proof The proof is an easy extension of that of Theorem 6.1; it is only necessary to apply Lemma 6.1 to the function $K\Phi + e$, where the constant K is chosen to ensure that the maximum principle applies. The details of the proof are left as an exercise. □

We now apply this theorem to the problem with the curved boundary, taking the set J_1 to contain all the ordinary internal mesh points, and J_2 to contain all those mesh points which have one or more neighbours on the boundary. We then define the mesh function Φ by

$$
\begin{aligned}
\Phi_P &= E_1 \left\{ (x_r - p)^2 + (y_s - q)^2 \right\} & \forall\, P \in J_\Omega, \\
\Phi_P &= E_1 \left\{ (x_r - p)^2 + (y_s - q)^2 \right\} + E_2 & \forall\, P \in J_{\partial\Omega},
\end{aligned}
$$

where E_1 and E_2 are positive constants to be chosen later. Then

$$
L_h \Phi_P = 4E_1 \quad \forall\, P \in J_1,
\tag{6.63a}
$$

but for points in J_2 there is an additional term, or terms, arising from the boundary points. In the approximation (6.33b) the coefficient of u_A is

$$
\frac{2}{\theta(1+\theta)(\Delta x)^2}
$$

and since $0 < \theta < 1$ this coefficient is bounded away from zero and cannot be less than $1/(\Delta x)^2$. Hence

$$
\begin{aligned}
L_h \Phi_P &\geq 4E_1 + E_2/(\Delta x)^2 \\
&\geq E_2/(\Delta x)^2 & \forall\, P \in J_2.
\end{aligned}
\tag{6.63b}
$$

Applying Theorem 6.2, with the truncation error bounds of (6.61) given by (6.57) as $T_1 = K_1(\Delta x)^2$ and $T_2 = K_2\Delta x$, we obtain

$$|e_P| \leq (E_1 R^2 + E_2) \max \left\{ \frac{K_1(\Delta x)^2}{4E_1}, \frac{K_2(\Delta x)^3}{E_2} \right\}. \tag{6.64}$$

This bound depends only on the ratio E_2/E_1, and is optimised when the two quantities in $\max\{\cdot, \cdot\}$ are equal; so we get the result

$$|e_P| \leq \tfrac{1}{4}K_1 R^2(\Delta x)^2 + K_2(\Delta x)^3, \tag{6.65}$$

showing that the error is in fact second order in the mesh size. Notice that the leading term in this error bound is unaffected by the lower order terms in the truncation error near the boundary.

As a second example, consider the solution of Poisson's equation in the unit square, with the Neumann boundary condition $u_x(1, y) = g(y)$ on the right-hand boundary $x = 1$, and Dirichlet conditions on the other three sides. As in a similar problem in Chapter 2 we introduce an extra line of points outside the boundary $x = 1$, with $r = J+1$. The boundary condition is approximated by

$$\frac{U_{J+1,s} - U_{J-1,s}}{2\Delta x} = g_s. \tag{6.66}$$

We then eliminate the extra unknown $U_{J+1,s}$ from the standard difference equation at $r = J$, giving

$$\frac{U_{J,s+1} + U_{J,s-1} + 2U_{J-1,s} - 4U_{J,s} + 2g_s\Delta x}{(\Delta x)^2} + f_{J,s} = 0. \tag{6.67}$$

This is now an equation of the general form (6.42), satisfying the required conditions for the maximum principle; so in the application of the maximum principle these points with $r = J$ are to be regarded as *internal* points.

The truncation error at the ordinary points is as before, and an expansion in Taylor series gives the truncation error of (6.67). The result is

$$T_{r,s} = \tfrac{1}{12}(\Delta x)^2(u_{xxxx} + u_{yyyy}) + O((\Delta x)^4), \quad r < J,$$
$$T_{J,s} = \tfrac{1}{12}(\Delta x)^2(u_{xxxx} + u_{yyyy}) - \tfrac{1}{3}\Delta x u_{xxx} + O\left((\Delta x)^3\right). \tag{6.68}$$

The same argument as before, using Theorem 6.1 and the comparison function Φ given in (6.13), shows that the error is bounded by

$$|e_{r,s}| \leq \tfrac{1}{8}\left\{ \tfrac{1}{12}(\Delta x)^2(M_{xxxx} + M_{yyyy}) + \tfrac{1}{3}\Delta x M_{xxx} \right\}. \tag{6.69}$$

This error bound is of order $O(\Delta x)$, but as in the previous example a sharper bound can be obtained by choice of a different comparison function and application of Theorem 6.2. We define

$$\Phi = (x-p)^2 + (y-q)^2 \tag{6.70}$$

where p and q are constants to be determined. We partition the internal points of the region into J_1 and J_2, where J_2 consists of those points with $r = J$. Then in region J_1 the standard difference equation is used, and

$$L_h\Phi_P = 4, \quad P \in J_1. \tag{6.71}$$

At the points where $r = J$, a different operator is used, as in (6.67), and we have

$$L_h\Phi_P = 4 - 4(1-p)/\Delta x, \quad P \in J_2. \tag{6.72}$$

In the notation of Theorem 6.2 we then find that

$$\begin{aligned}
\frac{T_1}{C_1} &= \frac{\frac{1}{12}(\Delta x)^2(M_{xxxx} + M_{yyyy})}{4}, \\
\frac{T_2}{C_2} &= \frac{\frac{1}{12}(\Delta x)^2(M_{xxxx} + M_{yyyy}) + \frac{1}{3}\Delta x M_{xxx}}{4 - (1-p)/\Delta x}.
\end{aligned} \tag{6.73}$$

If we now choose, for example, $p = 2, q = \frac{1}{2}$, we obtain

$$\frac{T_2}{C_2} \le \frac{1}{3}(\Delta x)^2 M_{xxx} + \frac{1}{12}(\Delta x)^3(M_{xxxx} + M_{yyyy}), \tag{6.74}$$

and at all points of the square we see that $(x-2)^2 + (y-\frac{1}{2})^2 \le \frac{17}{4}$; so by adding T_1/C_1 and T_2/C_2 we obtain the error bound

$$|e_{r,s}| \le \frac{17}{4}(\Delta x)^2[\frac{1}{3}M_{xxx} + \frac{1}{12}(\frac{1}{4} + \Delta x)(M_{xxxx} + M_{yyyy})]. \tag{6.75}$$

This shows that, in this example also, the error is second order in the mesh size.

The same technique can be used to show that the error in our approximation for the solution of Poisson's equation in a fairly general region, given either Dirichlet or Neumann conditions on a curved boundary, is still second order; the technique is complicated only by the need to take account of a number of different geometrical possibilities. Indeed, the same ideas can be applied quite generally to both elliptic and parabolic problems where maximum principles hold. Thus we end this section by sharpening some of the results that we gave in Chapter 2 for one-dimensional heat flow.

In Section 2.11 we gave a proof of convergence of the θ-method for the heat equation. We now define the operator L_h by

$$L_h\psi = \frac{[\theta\delta_x^2\psi_j^{n+1} + (1-\theta)\delta_x^2\psi_j^n]}{(\Delta x)^2} - \frac{\psi_j^{n+1} - \psi_j^n}{\Delta t}, \tag{6.76}$$

and take P as the point (x_j, t_{n+1}). It is easy to see that, provided $0 \le \theta \le 1$ and $\mu(1-\theta) \le \frac{1}{2}$, the conditions of Theorem 6.1 are satisfied for an appropriately defined set of points J_Ω. We shall suppose that Dirichlet conditions are given on $x = 0$ and $x = 1$; then the interior points J_Ω are those for which $1 \le j \le J-1$ and $1 \le n \le N$, and the boundary points $J_{\partial\Omega}$ have $j = 0$ or $j = J$ or $n = 0$. In the notation of Section 2.10, the exact solution u and the numerical solution U satisfy $L_h u_j^n = -T_j^{n+1/2}$ and $L_h U_j^n = 0$, so that

$$L_h e_j^n = T_j^{n+1/2}$$

and we have

$$T_j^{n+1/2} = O(\Delta t) + O((\Delta x)^2).$$

Now define the comparison function

$$\Phi_j^n = At_n + Bx_j(1 - x_j) \tag{6.77}$$

where A and B are nonnegative constants; it is easy to show that $L_h\Phi_j^n = -(A + 2B)$. Hence

$$L_h(e_j^n - \Phi_j^n) = A + 2B + T_j^{n+1/2}$$
$$\ge 0 \tag{6.78}$$

if the constants are so chosen that $A + 2B \ge T := \max |T_j^{n+1/2}|$.

At points on the boundary $J_{\partial\Omega}$ we have $e_j^n = 0$, $\Phi_0^n = \Phi_J^n = At_n$, and $\Phi_j^0 = Bx_j(1 - x_j)$, so that $e_j^n - \Phi_j^n \le 0$ on the boundary. Hence by Lemma 6.1, $e_j^n - \Phi_j^n \le 0$ in J_Ω. We now consider two choices of the constants A and B.

(i) Take $A = T$, $B = 0$; we have therefore shown that

$$e_j^n \le \Phi_j^n = t_n T, \tag{6.79}$$

and the same argument applied to the mesh function $(-e_j^n - \Phi_j^n)$ shows that

$$|e_j^n| \le t_n T \tag{6.80}$$

which agrees with (2.96).

(ii) Take $A = 0$, $B = \frac{1}{2}T$ and we obtain in the same way

$$|e_j^n| \le \tfrac{1}{2}x_j(1 - x_j)T. \tag{6.81}$$

From combining these results it is in fact clear that

$$|e_j^n| \le \max_{\substack{m<n \\ 0<i<J}}\{|T_i^{m+1/2}|\} \min\{t_n, \tfrac{1}{2}x_j(1 - x_j)\} \tag{6.82}$$

which properly reflects the fact that $e = 0$ round the boundary of the domain, and grows away from it.

We saw that for the model problem in Section 2.6 the error in the solution tends to zero as t increases, while the bound in (6.82) does not. If we have some additional information about the solution, and we can show that $|T_j^{n+1/2}| \le \tau_n$, where τ_n is a known decreasing function of n, we may be able to construct a comparison function which leads to an error bound which decreases with n. An example is given in Exercise 7.

In a similar way we can obtain an error bound in the case of a Neumann boundary condition, the problem considered in Section 2.13. With a homogeneous Neumann condition at $x = 0$, i.e. at x_0, the operator at $j = 1$ is replaced by

$$(L_h\psi)_1^n = \frac{\theta(\psi_2^{n+1} - \psi_1^{n+1}) + (1 - \theta)(\psi_2^n - \psi_1^n)}{(\Delta x)^2} - \frac{\psi_1^{n+1} - \psi_1^n}{\Delta t} \tag{6.83}$$

obtained from (2.103) with $\alpha = 0$ and $g = 0$. The truncation error, from (2.109), and with $\theta = 0$, at this point is

$$T_1^{n+1/2} = \tfrac{1}{2}\Delta t\, u_{tt} - \tfrac{1}{12}(\Delta x)^2 u_{xxxx} - \tfrac{1}{2}u_{xx}. \tag{6.84}$$

We now apply the argument of Theorem 6.2, with J_1 including all the interior mesh points with $j > 1$, and J_2 all the interior mesh points with $j = 1$. In (6.61) we can take $T_1 = T = \frac{1}{2}\Delta t\, M_{tt} + \frac{1}{12}(\Delta x)^2 M_{xxxx}$ as before, but in J_2 we must use $T_2 = T + \frac{1}{2}M_{xx}$. We then construct the comparison function such that $\Phi_0^n = \Phi_1^n$, namely

$$\Phi_j^n = \begin{cases} At_n + B(1 - x_j)(1 - \Delta x + x_j) & \text{in } J_1, \\ At_n + B(1 - x_j)(1 - \Delta x + x_j) + K & \text{in } J_2, \end{cases} \tag{6.85}$$

so that we obtain

$$L_h\Phi_j^n = \begin{cases} -(A + 2B) & \text{in } J_1, \\ -(A + 2B) - K/(\Delta x)^2 & \text{in } J_2. \end{cases} \tag{6.86}$$

This shows that $L_h(e^n_j - \Phi^n_j) \geq 0$ in J_1 and in J_2 if we choose the constants so that

$$A + 2B \geq T,$$
$$A + 2B + K/(\Delta x)^2 \geq T + \tfrac{1}{2}M_{xx}. \tag{6.87}$$

This is clearly satisfied if we take the same A and B as before and $K = \tfrac{1}{2}(\Delta x)^2 M_{xx}$.

The boundary of the region only involves the points where $j = J$ or $n = 0$, and at these points it is easy to see as before that $e^n_j - \Phi^n_j \leq 0$. Hence this holds at the interior points also, so that

$$e^n_j \leq At_n + B(1 - x_j)(1 - \Delta x + x_j) + \tfrac{1}{2}(\Delta x)^2 M_{xx}. \tag{6.88}$$

With the same choices of A and B as before we get the final result

$$|e^n_j| \leq \max_{\substack{m \leqslant n \\ 0 < i < J}} \{|T_i^{m+1/2}|\} \times$$

$$\min\{t_n, (1 - x_j)(1 - \Delta x + x_j) + \tfrac{1}{2}(\Delta x)^2 M_{xx}\}, \tag{6.89}$$

showing that the error is $O(\Delta t) + O((\Delta x)^2)$, just as in the case of Dirichlet boundary conditions.

6.6 Asymptotic error estimates

We had examples in the previous section where a straightforward error analysis gave a bound of first order in the mesh size, while a more sophisticated analysis produced a bound of second order. This must raise the question whether a still more careful analysis might show that the error is in fact of third order. In those examples it is fairly clear that no improvement in order is possible, but in more complicated problems it may not be at all easy to see what the actual order of error really is. For example, while in those examples the difficulties stemmed from lower order truncation errors at exceptional points near the boundary, the consideration of more general elliptic operators in Section 2.15, Section 3.5 and the next section can lead to lower order truncation errors at all points. It is therefore useful to have estimates which show more precisely how the error behaves in the limit as the mesh size tends to zero.

Such estimates often exploit more fully the maximum principle of Lemma 6.1, and the corresponding result which holds for an elliptic operator L. As a preliminary, suppose we denote by $\Phi_{\partial\Omega}$ the maximum of Φ_A over all the boundary nodes, as it appears in the bound of (6.52) with Φ satisfying (6.51). Now let us apply Lemma 6.1 to $\Psi := e - T(\Phi_{\partial\Omega} - \Phi)$: it is clear that

$$L_h \Psi_P = -T_P + T L_h \Phi_P \geq 0,$$

and also that the maximum of Ψ on the boundary is zero; so we can conclude that $\Psi_P \leq 0, \forall P \in J$. We can also repeat this argument with $-e$, and hence deduce that

$$|e_P| \leq T(\Phi_{\partial\Omega} - \Phi_P), \tag{6.90}$$

which can be a much stronger result than that of (6.52) in Theorem 6.1; in particular, it gives an error bound that decreases to zero at some point of the boundary.

To illustrate how to estimate the asymptotic behaviour of the error, we consider first the solution of Poisson's equation in the unit square, with Dirichlet conditions on the boundary. Using the standard five-point difference scheme we can easily write down an expression for the truncation error, and take more terms in the expansion because the underlying solution will be smooth for reasonable data:

$$\begin{aligned} T_{r,s} &= \tfrac{1}{12}(\Delta x)^2 (u_{xxxx} + u_{yyyy})_{r,s} \\ &+ \tfrac{1}{360}(\Delta x)^4 (u_{xxxxxx} + u_{yyyyyy})_{r,s} + \cdots. \end{aligned} \tag{6.91}$$

Then the error $e_{r,s}$ satisfies the equation

$$\begin{aligned} L_h e_{r,s} &= -T_{r,s} \\ &= -\tfrac{1}{12}(\Delta x)^2 (u_{xxxx} + u_{yyyy})_{r,s} + O\left((\Delta x)^4\right). \end{aligned} \tag{6.92}$$

Suppose now that we write $\psi(x,y)$ for the solution of the equation

$$\psi_{xx} + \psi_{yy} = -\tfrac{1}{12}(u_{xxxx} + u_{yyyy}) \tag{6.93}$$

which vanishes on the boundary of the unit square; and let Ψ be the result if we were to approximate this problem by our numerical scheme. Then an application of the error bound in Theorem 6.1 shows that $\Psi - \psi = O\left((\Delta x)^2\right)$. Moreover, we can combine this result with (6.92) so that another application of the theorem shows that

$$\frac{e_{r,s}}{(\Delta x)^2} = \psi(x_r, y_s) + O\left((\Delta x)^2\right). \tag{6.94}$$

That is, the numerical solution to the original Poisson's equation has an error which is given by

$$U_{r,s} = u(x_r, y_s) + (\Delta x)^2 \psi(x_r, y_s) + O\left((\Delta x)^4\right). \tag{6.95}$$

This shows that the error is exactly second order, and not of any higher order, except in a very special case where the function ψ is identically zero; but of course the expression is only valid in the limit as the mesh size goes to zero.

We can estimate the size of this error from the difference between two numerical solutions using different mesh sizes, comparing the results at the mesh points they have in common. Alternatively, we could use divided differences of the approximate solution U to estimate the right-hand side of (6.93) and then actually solve the discrete equations again to obtain an approximation to what we called ψ above; then substitution into (6.95) leads to a fourth order approximation. This procedure is called *deferred correction*. Again we should emphasise that the extra orders of accuracy will only be obtained if the mesh size is sufficiently small for the asymptotic expansion (6.95) to be valid.

An asymptotic estimate can be obtained in a similar way for the last example in the previous section, with a Neumann boundary condition on one side of the square, where the error bound was obtained by a rather arbitrary choice of the comparison function Φ; it is clear that the error bound (6.75) is most unlikely to be the best possible. In this case the error satisfies $L_h(e_{rs}) = -T_{rs}$ where T_{rs} is given by (6.68). If we now define ψ to be the solution of the problem

$$\psi_{xx} + \psi_{yy} = -\tfrac{1}{12}(u_{xxxx} + u_{yyyy}) \qquad (6.96)$$

with

$$\psi(0, y) = \psi(x, 0) = \psi(x, 1) = 0, \quad \psi_x(1, y) = -\tfrac{1}{6}u_{xxx}(1, y), \qquad (6.97)$$

we find that the same numerical method applied to this problem will lead to the equations satisfied by $e_{r,s}$, with additional truncation terms of higher order; the details are left as an exercise – see Exercise 6. We thus see, as in the previous example, that

$$U_{r,s} = u(x_r, y_s) + (\Delta x)^2 \psi(x_r, y_s) + o((\Delta x)^2), \qquad (6.98)$$

and $\psi(.,.)$ could be estimated by the deferred correction approach.

The extension to problems involving a curved boundary is straightforward. As in the application immediately following Theorem 6.2 we divide the set of mesh points into J_1 and J_2, where J_2 comprises those mesh points that have one or more neighbours on the boundary. At points in J_1 we have, as in (6.92),

$$\left| L_h(e_{rs}) + \tfrac{1}{12}(\Delta x)^2(u_{xxxx} + u_{yyyy}) \right| \leq K_3 (\Delta x)^4 \qquad \text{in } J_1, \qquad (6.99)$$

where $K_3 = \frac{1}{360}(M_{xxxx} + M_{yyyy})$. We now suppose that Dirichlet conditions are given on the curved boundary. Then at points in J_2 we can use (6.56) and (6.57b) to give

$$\left|L_h(e_{rs})\right| \leq K_2 \Delta x \qquad \text{in } J_2. \tag{6.100}$$

We define $\psi(x, y)$ as in (6.93) to be the solution of

$$\psi_{xx} + \psi_{yy} = -\tfrac{1}{12}\left(u_{xxxx} + u_{yyyy}\right) \tag{6.101}$$

which vanishes on the boundary. Then, provided the functions u and ψ have sufficient bounded derivatives, there exist constants K_4 and K_5 such that

$$\left|L_h(\psi_{rs}) + \tfrac{1}{12}(u_{xxxx} + u_{yyyy})\right| \leq K_4 (\Delta x)^2 \qquad \text{in } J_1$$
$$\left|L_h(\psi_{rs})\right| \leq K_5 \Delta x \qquad \text{in } J_2. \tag{6.102}$$

By subtracting (6.102) from (6.99) and (6.100) we then have

$$\left|L_h\left(\frac{e_{rs}}{(\Delta x)^2} - \psi_{rs}\right)\right| \leq (K_3 + K_4)(\Delta x)^2 \qquad \text{in } J_1,$$
$$\left|L_h\left(\frac{e_{rs}}{(\Delta x)^2} - \psi_{rs}\right)\right| \leq \frac{K_2}{\Delta x} + K_5 \Delta x \qquad \text{in } J_2. \tag{6.103}$$

We now apply Theorem 6.2; the function Φ_P and the values of C_1 and C_2 are the same as those used to obtain (6.64), i.e., $C_1 = 4E_1$, $C_2 = E_2/(\Delta x)^2$, but $T_1 = (K_3 + K_4)(\Delta x)^2$ and $T_2 = K_2/\Delta x + K_5 \Delta x$. We thus obtain

$$\left|\frac{e_{rs}}{(\Delta x)^2} - \psi_{rs}\right| \leq (E_1 R^2 + E_2) \max\left\{\frac{(K_3 + K_4)(\Delta x)^2}{4E_1}, \frac{K_2 + K_5(\Delta x)^2}{E_2/\Delta x}\right\}. \tag{6.104}$$

Choosing $E_1 = \tfrac{1}{4}(K_3 + K_4)(\Delta x)^2$, $E_2 = K_2 \Delta x + K_5(\Delta x)^3$ we then obtain

$$\left|\frac{e_{rs}}{(\Delta x)^2} - \psi_{rs}\right| \leq C(\Delta x), \tag{6.105}$$

showing that

$$e_{rs} = (\Delta x)^2 \psi(x_r, y_s) + O\big((\Delta x)^3\big). \tag{6.106}$$

Once again this shows how the lower order truncation error at points near the boundary does not affect the leading term in the asymptotic expansion of the error.

A combination of the asymptotic error estimates obtained in this section and the rigorous error bounds of previous sections will give a useful description of the behaviour of the error in the linear problems with smooth solutions that we have been discussing. We should note, however, that for the more general equation involving a variable $a(x, y)$ the construction of a function Φ with the required properties is likely to be more difficult; just how difficult, and how sharp will be the resulting bound, will depend on the form of $a(x, y)$, a lower bound for which may well have to be used in the construction.

However, an attempt to apply the analysis given above to a problem with Neumann conditions on part of a curved boundary is less successful. If we use the approximation of (6.41) and expand in Taylor series we find the leading term of the truncation error is

$$\frac{p_1^2 - p_2^2}{1 + p_1 + p_2} u_{xy}. \tag{6.107}$$

We see at once that this is $O(1)$, and does not tend to zero as the mesh size goes to zero. It is now not difficult to use the maximum principle to obtain a bound on the error, and we find a bound which is $O(\Delta x)$. But our asymptotic expansion of the error is founded on the fact that the leading term in the truncation error is the product of a power of Δx and a function of x and y only, independent of Δx. This is no longer true of the expression in (6.107), since p_1 and p_2 are functions of x, y and Δx, and moreover are not smooth functions, but involve terms like the fractional part of $x/\Delta x$. This observation is illustrated in Fig. 6.4. This shows the results of a numerical solution of Poisson's equation on the annulus between circles of radii 1 and 0.3. In the first problem Dirichlet conditions are given on both boundaries, and these conditions and the function f are so chosen that the solution is

$$u(x, y) = 1 + 3x^2 + 5y^2 + 7x^2y^2 + (x^2 + 2)^2.$$

The maximum error in this calculation is shown in the lower curve, which shows the error behaving like $(\Delta x)^2$. In the second problem the solution is the same, but a Neumann condition is given on the outer circular boundary. We notice that the error is now much larger, that there is a general trend of order $O(\Delta x)$, but the detailed behaviour is very irregular. To get a numerical solution of such a problem, with an error behaving smoothly like $O((\Delta x)^2)$, will require a more complicated approximation to the boundary condition.

A major advantage of the finite element method, as we shall see in the next section, is that it deals with the application of Neumann boundary conditions in a simple and natural way, that leads to much better behaved errors.

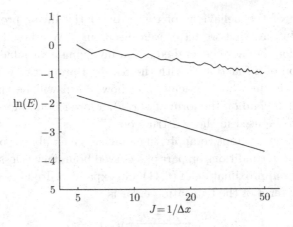

Fig. 6.4. Numerical solution of Poisson's equation in an annulus: lower curve, Dirichlet conditions; upper curve, Neumann condition on the outer boundary.

6.7 Variational formulation and the finite element method

Our general diffusion problem (6.20), (6.22) can be given a variational formulation. Consider first the case where Dirichlet boundary conditions are given at all points on the boundary. We define

$$I(v) := \int_\Omega \left[\tfrac{1}{2} a |\nabla v|^2 - fv \right] \mathrm{d}x\, \mathrm{d}y. \tag{6.108}$$

Then we assert that the solution of (6.20) satisfies a variational equation that we write as

$$\delta I(u) = 0. \tag{6.109}$$

This means that if $v = u + \delta u$ is any function for which $I(v)$ is defined, and which satisfies the boundary conditions, then

$$I(u + \delta u) - I(u) = o(\delta u). \tag{6.110}$$

We can show why this is so without attempting a completely rigorous proof. By expanding $I(u + \delta u)$ we find that

$$I(u + \delta u) - I(u) = \int_\Omega \left[(a\nabla u) \cdot (\nabla \delta u) - f \delta u \right] \mathrm{d}x\, \mathrm{d}y + \int_\Omega \tfrac{1}{2} a |\nabla \delta u|^2 \mathrm{d}x\, \mathrm{d}y$$

$$= \int_\Omega - \left[\nabla \cdot (a\nabla u) + f \right] \delta u\, \mathrm{d}x\, \mathrm{d}y + O\left((\delta u)^2 \right), \tag{6.111}$$

where we have used Gauss' theorem, and the fact that $\delta u = 0$ on the boundary of Ω. The result follows.

In fact we have shown rather more, for since $a(x, y) > 0$ it follows from (6.111) that $I(u + \delta u) \geq I(u)$ for all functions δu which vanish on the boundary. Hence the function u gives the minimum of $I(v)$, taken over all functions v which satisfy the boundary conditions.

Now suppose we take a finite expansion of the form

$$V(x, y) = \sum_{j=1}^{N} V_j \phi_j(x, y), \tag{6.112}$$

where the functions ϕ_j are given, and try to choose the coefficients V_j so that this gives a good approximation to the solution u. We can expand $I(V)$ in the form

$$I(V) = \tfrac{1}{2} \sum_i \sum_j A_{ij} V_i V_j - \sum_i b_i V_i, \tag{6.113}$$

where

$$A_{ij} = \int_\Omega [a \nabla \phi_i \cdot \nabla \phi_j] \mathrm{d}x \, \mathrm{d}y \tag{6.114}$$

and

$$b_i = \int_\Omega f \phi_i \mathrm{d}x \, \mathrm{d}y. \tag{6.115}$$

Since the exact solution u minimises $I(v)$, it is natural to define an approximation U to the solution by choosing the coefficients $\{V_j\}$ to minimise $I(V)$. As (6.113) shows that $I(V)$ is a quadratic form in the coefficients, the determination of its minimum is a straightforward matter, applying the constraint that V satisfies the boundary conditions.

This approach was used by Rayleigh, Ritz and others in the nineteenth century with various choices of the functions $\phi_j(x, y)$. It is also the starting point for the finite element method which is now widely used to solve elliptic problems in preference to finite difference methods, especially by engineers.

The particular feature of the finite element method lies in the choice of the $\phi_j(x, y)$, which are known as the *trial functions*, or *shape functions*. They are chosen so that each ϕ_j is non-zero only in a small part of the region Ω. In the matrix A most of the elements will then be zero, since A_{ij} will only be non-zero if the shape functions ϕ_i and ϕ_j overlap. We will consider one simple case.

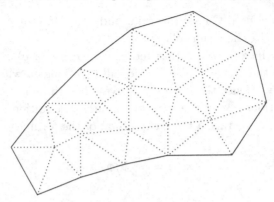

Fig. 6.5. Triangular elements in a polygon.

Suppose the region Ω is a polygon and it is subdivided into triangular elements, as in Fig. 6.5. The vertices P of the triangles are known as *nodes*. We suppose that V is piecewise linear, having a linear form in each triangle determined by its values at the vertices of the triangle. Then it is clearly continuous from one triangle to the next; and we suppose that the boundary conditions are similarly piecewise linear around the bounding polygon and V takes on these values. The function $\phi_j(x, y)$ is defined to be the piecewise linear function which takes the value 1 at node P_j, and zero at all the other nodes. This defines the function uniquely, and it is clear that it is non-zero only in the triangles which have P_j as a vertex. This function is for obvious reasons known as a *hat function*; drawn in three dimensions (Fig. 6.6) it has the form of a pyramid, with triangular faces.

This definition of the shape functions has the useful property that the coefficient V_j gives the value of the function $V(x, y)$ at the node P_j, since all the other shape functions vanish at this node. Then to ensure that V satisfies the boundary conditions we fix the coefficients V_j corresponding to nodes on the boundary, and allow all the others to vary in the minimisation process.

One of the main advantages of the finite element method is that it adapts quite easily to difficult geometrical shapes, as a triangulation can easily follow a boundary of almost any shape. Once a set of triangular elements has been constructed, the elements of the matrix A can be evaluated in a routine way, taking no particular account of the shape of the triangles.

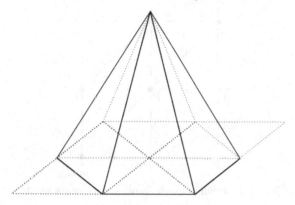

Fig. 6.6. A hat basis function.

To illustrate the procedure we shall consider the simplest model problem, the solution of Poisson's equation in a square with $u = 0$ on the boundary, as in (6.1). We cover the square with a uniform square mesh as before, and then divide each small square into two triangles by drawing a diagonal, as in Fig. 6.7. A general row of the matrix A will then contain seven non-zero elements, as the hat function centred at the node P will overlap with the hat functions centred at P and its six neighbours, labelled N, S, E, W, NW and SE in the figure. Since each ϕ is a linear function of x and y in each triangle, its gradient $\nabla \phi$ is constant in each triangle. With $a(x, y) = 1$ in this example, the evaluation of

$$A_{ij} = \int_\Omega [\nabla \phi_i \cdot \nabla \phi_j] \mathrm{d}x \, \mathrm{d}y \qquad (6.116)$$

is a simple matter. The partial derivatives of ϕ in each triangle each take one of the values 0, $1/\Delta x$ or $-1/\Delta x$, and we find that

$$A_{PP} = 4, \qquad (6.117a)$$

$$A_{PN} = A_{PS} = A_{PE} = A_{PW} = -1 \qquad (6.117b)$$

and

$$A_{P,NW} = A_{P,SE} = 0. \qquad (6.117c)$$

We also require the value of b_i from (6.115). If we approximate this by replacing the function $f(x, y)$ by a constant, the value of f at the point P, we obtain

$$b_P = (\Delta x)^2 f_P. \qquad (6.118)$$

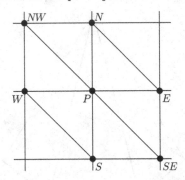

Fig. 6.7. Triangular elements on a square mesh.

However, we should point out here that in a finite element program the integrations of (6.114) and (6.115) will be carried out over each element before being assembled into the global equation of the form (6.119) below: in particular, this will mean that (6.115) will commonly be approximated by centroid quadrature in each triangle, so that in (6.118) f_P will be replaced by the mean of the six values at the centroids of the triangles sharing the vertex P. Now the minimum of the quadratic form (6.113) is given by the vector $\mathbf{U}$, with values that satisfy the system of linear equations $A\mathbf{U} = \mathbf{b}$. The number of interior nodes corresponding to the V_j which are allowed to vary, determines the dimension of $\mathbf{U}$, $\mathbf{b}$ and hence of A; there are no other contributions to the right-hand side of the equations because the boundary data are zero. A general equation of this system is

$$4U_P - U_N - U_S - U_E - U_W - (\Delta x)^2 f_P = 0, \qquad (6.119)$$

which is the same as the familiar finite difference scheme introduced in Section 6.1 – apart from the change of sign throughout.

The error analysis of the finite element method has a different character from that of the finite difference schemes. We have seen that u satisfies

$$\int_\Omega [a\nabla u \cdot \nabla w - f\,w]\,\mathrm{d}x\,\mathrm{d}y = 0 \qquad (6.120)$$

for any function w which vanishes on the boundary. Also the function U would satisfy

$$\int_\Omega [a\nabla U \cdot \nabla W - f\,W]\,\mathrm{d}x\,\mathrm{d}y = 0 \qquad (6.121)$$

if the integrals were carried out exactly (or a and f were constant), for any function W which can be expressed as a finite sum of the form (6.112) and vanishes on the boundary. Now we can choose any V of the form (6.112) which satisfies the boundary conditions and take both w and W to be $V - U$ because this difference vanishes on the boundary; then by subtraction we obtain

$$\int_\Omega [a\nabla(U - u) \cdot \nabla(V - U)]dx\,dy = 0. \tag{6.122}$$

Thus

$$\int_\Omega a|\nabla(V - u)|^2 dx\,dy = \int_\Omega a|\nabla[(V - U) + (U - u)]|^2 dx\,dy$$

$$= \int_\Omega a|\nabla(V - U)|^2 dx\,dy + \int_\Omega a|\nabla(U - u)|^2 dx\,dy,$$

because the cross-product terms drop out by (6.122). This means that

$$\int_\Omega a|\nabla(U - u)|^2 dx\,dy \leq \int_\Omega a|\nabla(V - u)|^2 dx\,dy \quad \forall V. \tag{6.123}$$

We can define a special norm for a function $w(x, y)$ which vanishes on the boundary of Ω,

$$||w||_E^2 = \int_\Omega a|\nabla w|^2 dx\,dy. \tag{6.124}$$

We have thus shown that

$$||U - u||_E \leq ||V - u||_E \quad \forall V. \tag{6.125}$$

This key result means that U is the best possible approximation to u of the form (6.112) in this norm. Its error can then be estimated very sharply by application of approximation theory. To do so here in any detail would take us beyond the scope of this book, as would consideration of the effects of the quadrature needed for variable a and f; we merely quote the main results as

$$||U - u||_E \leq C_1 h|u|_*, \tag{6.126a}$$

$$||U - u||_{L_2(\Omega)} \leq C_2 h^2 |u|_*, \tag{6.126b}$$

where h is the maximum diameter of all the triangles in the triangulation, $|u|_* = ||u_{xx}|| + ||u_{xy}|| + ||u_{yy}||$ and $||\cdot||$ denotes the $||\cdot||_{L_2(\Omega)}$ norm.

Finally, suppose that homogeneous Neumann boundary conditions are imposed on part of the boundary; these are dealt with as a natural part of the variational process and in a correspondingly automatic way by the finite element method. In the derivation of (6.111) a boundary integral

$$\int_{\partial\Omega} a\frac{\partial u}{\partial n}\delta u\, dl \tag{6.127}$$

is obtained from the application of Gauss' theorem. This is zero if at all boundary points either Dirichlet boundary conditions are imposed, giving $\delta u = 0$, or a homogeneous Neumann boundary condition is required, in which case $\partial u/\partial n = 0$ for the true solution. Thus when we minimise $I(v)$ given by (6.108), at the minimum not only is (6.20) satisfied at all interior points, but also $\partial u/\partial n = 0$ at all boundary points which are not constrained by a Dirichlet boundary condition. This is called a *natural* treatment of the Neumann boundary condition.

The finite element method makes direct use of this minimisation property. Boundary nodes which are not constrained by a Dirichlet boundary condition are treated as interior nodes and included in the sum (6.112), with associated shape functions which are truncated by the boundary. Thus the computation of the system of equations $A\mathbf{U} = \mathbf{b}$ can be performed as before, except that its dimension will be increased because of the additional nodes and the corresponding entries in A and $\mathbf{b}$ will be of non-standard form – see the example in Section 6.9. Moreover, the error analysis outlined above will continue to apply; the only changes required are to replace the phrases 'satisfy the boundary condition' and 'vanish on the boundary' by 'satisfy the Dirichlet boundary condition' and 'vanish on the Dirichlet boundary'.

6.8 Convection–diffusion problems

As we have seen in earlier chapters, the addition of lower order terms to a diffusion operator can cause difficulties and a loss of performance with many standard methods. The differential equation (6.24) that we have already discussed in this chapter still satisfies a maximum principle; this is because if there were an interior maximum, at that point we would have $u_x = u_y = 0$ so that these terms do not affect the deduction by contradiction that there can be no such maximum. (Note, moreover, this would be true even if these terms arose independently of the diffusion coefficient a.) But the corresponding argument in Lemma 6.1 for the discrete equation (6.25) breaks down unless the condition (6.27) is satisfied, and this can be very restrictive. Fortunately, in the case discussed

there this difficulty can be avoided by using the finite volume approach
to deriving the difference equations, which yields the scheme (6.29a).

It is in situations where the first order terms do not arise from expanding the diffusion term that the difficulties become much more serious.
The distinction is highlighted by the variational form just considered
in Section 6.7: from the self-adjoint differential operator of (6.20) we
obtain in (6.120) the bilinear form

$$\int_\Omega a\nabla u \cdot \nabla w \, dx \, dy \qquad (6.128)$$

which is symmetric in u and w and hence leads to the norm defined
in (6.124). But suppose we have a *convection–diffusion* equation of the
form

$$-\nabla(\epsilon\nabla u) + \mathbf{V} \cdot \nabla u = f. \qquad (6.129)$$

Here ϵ is a (positive) diffusion coefficient, the extra term compared with
equation (6.20) comes from a convective (or advective) velocity $\mathbf{V}$, and
we have introduced a change of notation so as to emphasise the fact that
the diffusion coefficient is often quite small. If the velocity is incompressible, that is $\nabla \cdot \mathbf{V} = 0$, this equation is equivalent to

$$\nabla \cdot (\epsilon\nabla u - \mathbf{V}u) + f = 0. \qquad (6.130)$$

Either form may occur in practical problems and they correspond to a
steady, two-dimensional version of the problem considered in Section 5.7;
and they raise the same issues. By multiplying the latter equation by
a *test function* w, integrating over the domain and integrating by parts
we are led to the bilinear form

$$\int_\Omega (\epsilon\nabla u - \mathbf{V}u) \cdot \nabla w \, dx \, dy. \qquad (6.131)$$

This is now no longer symmetric in u and w. As a result we do not have
a natural norm in which to measure the accuracy of an approximation,
and the system of equations corresponding to (6.121) no longer gives the
best approximation to u.

We will consider briefly below how finite difference methods should
be derived to restore the maximum principle, and how finite element
methods can come close to restoring the best approximation property.

Finite difference schemes are best formulated through the finite volume approach, combining the formulas of (6.29) for the diffusion terms
with those derived in Section 4.7 for the convective terms. In the latter case we found it desirable to use some form of upwinding: thus if

the velocity component $V^{(x)}$ is positive, in the compass point notation of (6.30) we replace the east and west fluxes by the following approximations:-

$$\left(\epsilon \frac{\partial u}{\partial x} - V^{(x)} u \right)_e \approx \epsilon_e \frac{U_E - U_P}{\Delta x} - V_P^{(x)} U_P$$

$$\left(\epsilon \frac{\partial u}{\partial x} - V^{(x)} u \right)_w \approx \epsilon_w \frac{U_P - U_W}{\Delta x} - V_W^{(x)} U_W. \qquad (6.132)$$

If the velocity $V^{(y)}$ is also positive and we make a similar replacement for the north and south fluxes, the resultant replacement equation for (6.30) satisfies all the conditions of Lemma 6.1 and the approximation satisfies a maximum principle.

We regard this as the simplest upwind choice because, although there is no time derivative in equation (6.130), the sign of ϵ requires that if it were introduced it would be on the right-hand side: thus the characteristics would point from bottom left to top right. The penalty to be paid for such a simple way of ensuring a maximum principle holds is that we have only first order accuracy. Convection–diffusion problems often exhibit steep boundary layers, and this scheme, while ensuring that a monotone layer retains that property, will often result in a much thicker layer than the correct result. This effect, and an estimate of its magnitude, are easily obtained by a modified equation analysis which will show that the truncation error from the first order approximation of the convection term will enhance the diffusion term.

In a finite element approximation of (6.130) we will also introduce some upwinding. This is best done by modifying the test functions that are used. When, as in (6.121), the same basis functions ϕ_j are used to generate the test functions W as are used in the expansion (6.112) for the approximation U the result is called a *Galerkin method*. Use of more general test functions results in what is called a *Petrov–Galerkin method*. We consider useful choices for a convection–diffusion problem, firstly in one dimension.

The piecewise linear basis function, depicted in Fig. 6.6, is given in one dimension by

$$\phi_j(x) = \begin{cases} (x - x_{j-1})/h_j & x \in e_j, \\ (x_{j+1} - x)/h_{j+1} & x \in e_{j+1}, \end{cases} \qquad (6.133)$$

where the element e_j of length h_j corresponds to the interval (x_{j-1}, x_j). Now suppose we introduce an upwinded test function as

$$\psi_j(x) = \phi_j(x) + \alpha \sigma_j(x), \tag{6.134}$$

where we define the modifying $\sigma_j(\cdot)$ as

$$\sigma_j(x) = \begin{cases} 3(x - x_{j-1})(x_j - x)/h_j^2 & x \in e_j, \\ -3(x - x_j)(x_{j+1} - x)/h_{j+1}^2 & x \in e_{j+1}. \end{cases} \tag{6.135}$$

Thus a quadratic bubble is added to $\phi_j(\cdot)$ to the left of x_j, if α is positive, and a similar bubble subtracted on the right. Now let us apply this to the equation that corresponds to (6.121),

$$\int [\epsilon U' - VU] \psi_j' \, dx = \int f \psi_j \, dx. \tag{6.136}$$

For simplicity we assume ϵ and V are constants. Then, because U' is constant on each interval and $\sigma_j(\cdot)$ is zero at each node, the upwinding does not affect the diffusion term and we get the scheme

$$-\epsilon \left[(U_{j+1} - U_j)/h_{j+1} - (U_j - U_{j-1})/h_j \right]$$
$$+ V \left[\tfrac{1}{2}(1 - \alpha)(U_{j+1} - U_j) + \tfrac{1}{2}(1 + \alpha)(U_j - U_{j-1}) \right] = \int f \psi_j \, dx. \tag{6.137}$$

It is clear that when V is positive taking $\alpha > 0$ corresponds to upwinding. Moreover, in that case and with a uniform mesh spacing, the scheme satisfies a maximum principle if

$$\tfrac{1}{2}(1 - \alpha)V \le \epsilon/h, \quad \text{i.e.,} \quad \alpha \ge 1 - 2\epsilon/Vh. \tag{6.138}$$

Note too that we can write (6.137) on a uniform mesh as

$$-(\epsilon/h + \tfrac{1}{2}\alpha V)\delta_x^2 U_j + V \Delta_{0x} U_j = \int f \psi_j \, dx, \tag{6.139}$$

which shows how upwinding increases the diffusion. As we have seen before, when the mesh Péclet number Vh/ϵ is less than 2 we can take $\alpha = 0$ and use central differencing for the convective term; but for any larger value we can choose α, for example by taking the equality sign in (6.138), to ensure a maximum principle holds.

A wide variety of upwinding 'bubbles' can be (and have been) used to achieve the above results. This provides many possibilities for generalising the techniques to more dimensions. A good indicator of where the bubble should be placed is provided by $\mathbf{V} \cdot \nabla \phi_j$: indeed, this forms the starting point for the very widely used *streamline diffusion* methods. We refer the reader to the Bibliographic notes for references to these and other methods. In these references one will also find that, in various norms, there are choices of test functions which will yield approximations that are within a given factor of best possible.

6.9 An example

As an example to conclude this chapter we consider a problem whose geometry is similar to a problem considered in Chapter 3 – see Fig. 3.6. We wish to determine the electrical capacity of a conductor consisting of two long cylinders, the inner one being circular, and the outer one being square in cross-section. This leads to a two-dimensional problem, requiring the solution of Laplace's equation in the region between a square and a concentric smaller circle. This problem has three lines of symmetry, and it is clearly enough, and much more efficient, to consider only one eighth of the region, as shown in Fig. 6.8.

We need to solve Laplace's equation in the region bounded by the circular arc AD and the lines AB, BC and CD. The boundary conditions specify that the solution is zero on the circular arc AD, and is equal to unity on the line BC. The other two lines are symmetry boundaries; formally we could specify that the normal derivative vanishes on these two lines, but a simpler and more accurate approach uses symmetry directly, by first ensuring that these boundary lines pass through mesh points and then treating these points as interior points with non-standard equations.

When using a finite difference scheme, at a point P on the symmetry boundary AB we construct a neighbouring mesh point S outside the boundary. By symmetry, $U_S = U_N$, so the difference equation corresponding to the point P becomes

$$c_N U_N + c_S U_N + c_E U_E + c_A U_A - c_P U_P = 0. \qquad (6.140)$$

Similarly on the symmetry line CD we construct points outside the boundary at N' and W', so that the use of symmetry with

$$c_{N'} U_{N'} + c_{S'} U_{S'} + c_{E'} U_{E'} + c_{W'} U_{W'} - c_{P'} U_{P'} = 0 \qquad (6.141)$$

produces

$$(c_{W'} + c_{S'}) U_{S'} + (c_{E'} + c_{N'}) U_{E'} - c_{P'} U_{P'} = 0. \qquad (6.142)$$

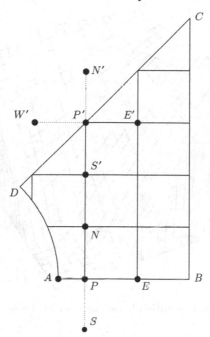

Fig. 6.8. Example: Laplace's equation.

The standard difference scheme is used at all points except those adjacent to the circular arc, and there the coefficients are calculated as discussed in Section 6.4.

When applying the finite element method to this problem, the values at the nodes on the interior of the edges AB and CD are included in the minimisation process. Each has an associated shape function which is truncated at the boundary, and so yields a modified equation. We suppose the diagonals for the triangular elements are drawn in the direction CD. Then for the typical point P' on CD it is clear from the symmetry that the coefficient $c_{P'}$ is halved and $c_{E'}$ and $c_{S'}$ are unchanged by the truncation of the shape functions; thus we get equation (6.142) multiplied by a half. At a point P on AB the coefficients c_P, c_E and c_W are all halved and c_N is unchanged; so again we obtain (6.140) multiplied by a half.

Thus in this case, the finite difference and finite element approaches give the same system of equations. However, the latter clearly yields a more general procedure for dealing with Neumann boundary conditions, which is much more straightforward than that described in Section 6.4.

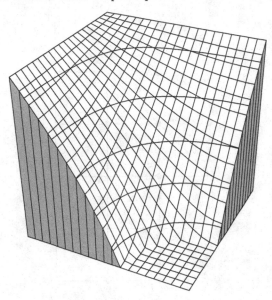

Fig. 6.9. Solution in one quarter of the region.

A solution obtained with a unit square and a circle of radius 0.4 is illustrated in Fig. 6.9. The solution has been reflected in the diagonal line for ease of visualisation.

The result required for this problem is the flux of the electric field out of the boundary. This is given by the flux either across the circular arc AD, or across the straight line BC. The latter is obviously easier to calculate accurately; as we are only considering one eighth of the region, we therefore need to compute

$$g = 8 \int_{BC} \frac{\partial u}{\partial x} \, dy. \tag{6.143}$$

Using a central difference, this is approximated at the point (x_r, y_s) by

$$\frac{U_{r+1,s} - U_{r-1,s}}{2\Delta x}$$

with one of the points being outside the boundary. Using the standard five-point approximation to Laplace's equation at the boundary point we can eliminate $U_{r+1,s}$ to give the approximation

$$\frac{4U_{r,s} - 2U_{r-1,s} - U_{r,s+1} - U_{r,s-1}}{2\Delta x}.$$

Table 6.1. *Calculation of flux for various* Δx.

Δx	g
0.1000	6.33901
0.0500	6.33495
0.0250	6.33396
0.0125	6.33370

With this approximation to the derivative, except at the point C where $\partial u/\partial n$ is assumed to be zero, the required flux is easily calculated from (6.143) by the use of the trapezium rule.

It is left as an exercise for the reader to check that the same formula would be obtained from a finite element integral of the form

$$\int_\Omega [\nabla U \cdot \nabla W] \, dx \, dy, \tag{6.144}$$

evaluated in the same way as the integrals in (6.116), where W is the piecewise linear function which has unit value on BC and is zero on the vertical line through E.

Numerical results for various mesh sizes are shown in Table 6.1. These clearly indicate an error of order $(\Delta x)^2$, with a correct value of 6.33362. In fact the numerical results are fitted to four decimal places by the expression

$$6.33362 + 0.53(\Delta x)^2.$$

Bibliographic notes and recommended reading

An analysis of methods for elliptic problems with general boundary conditions was given in the classic text by Collatz (1966) already referred to. The idea of the construction of difference equations by integrating over small elements of area seems to have been widely publicised first in Varga (1974).

The use of a maximum principle in the error analysis of finite difference methods was developed by Bramble and Hubbard in a series of papers beginning in 1964, a selection of which are referenced in the bibliography. For a general account of the use of maximum principles for elliptic equations, see the text of Protter and Weinberger (1967).

Our brief introduction to finite element methods does scant justice to the impact that they have had on the solution of elliptic problems, which has been covered in many books – see, for example, Strang and Fix (1973), Ciarlet (1978), and Brenner and Scott (2002). The many further aspects covered in such books include higher order shape functions, quadrilateral and curved elements and error analysis in various norms.

A key feature of finite element methods is the possibility of locally refining the mesh or introducing higher order shape functions in selected elements. Such local refinement in selected elements, in which the error is worse than the average, is based on an *a posteriori* error estimate, that is an estimate of the error derived from the computed approximation. A simple introduction to these ideas can be found in Süli and Mayers (2003).

The special difficulties encountered in convection–diffusion problems (both with finite volume and finite element methods) can be found in the books by Roos *et al.* (1996) and Morton (1996).

Exercises

6.1   To obtain a numerical solution of Poisson's equation

$$u_{xx} + u_{yy} + f(x,y) = 0$$

on the triangle bounded by the lines $y = 0$, $y = 2 + 2x$ and $y = 2 - 2x$, with Dirichlet conditions given at all points on the boundary, we use a uniform square grid of size $\Delta x = \Delta y = 1/N$. Find the leading terms in the truncation error of the standard five-point difference schemes at internal points, and at points adjacent to the boundary. Show how to choose the constant C so that a maximum principle may be applied to the mesh function $u(x_i, y_j) - U_{i,j} + Cy_j^2$. Deduce that the error in the solution is at least first order in the mesh size.

The necessity for a special difference scheme near the boundary could be avoided by using a rectangular mesh with $\Delta y = 2\Delta x$. Would this give a real advantage over the previous method?

6.2   We require the solution of the equation

$$\frac{\partial}{\partial x}\left(a\frac{\partial u}{\partial x}\right) + \frac{\partial}{\partial y}\left(a\frac{\partial u}{\partial y}\right) + f(x,y) = 0$$

in the unit square, given Dirichlet conditions at all points on the boundary. The difference scheme

$$L_h U := \frac{\delta_x(a\delta_x U)}{(\Delta x)^2} + \frac{\delta_y(a\delta_y U)}{(\Delta y)^2} = -f$$

is used, with a uniform grid. Suppose that a is a positive monotonic increasing function of x only; show how to choose the constant C so that the comparison function $\Phi = C(x^2 + y^2)$ satisfies the conditions of Theorem 6.1. Extend this result to the cases where (i) a is a positive monotonic decreasing function of x only, and (ii) $a(x,y)$ is positive, monotonic increasing as a function of x for fixed y, and monotonic decreasing as a function of y for fixed x.

6.3   Construct explicitly the system of linear equations obtained from approximating Poisson's equation $u_{xx} + u_{yy} + f(x,y) = 0$ in the region defined by $x \geq 0$, $y \geq 0$, $x^2 + y \leq 1$. Use a uniform square grid of size $\frac{1}{3}$. The boundary conditions are $u(x,0) = p(x)$, $u_x(0,y) = q(y)$, and $u(x, 1 - x^2) = r(x)$, where p, q, r and f are given functions.

6.4   The central difference approximation is used on a uniform mesh of size $\Delta x = a/(M+1)$, $\Delta y = b/(N+1)$ to solve the equation

$$u_{xx} + u_{yy} - Ku + f(x,y) = 0,$$

on the rectangle $0 \leq x \leq a$, $0 \leq y \leq b$, where K is a positive constant, and f is a given function; Dirichlet conditions are given at all points on the boundary.

Write down the leading terms in the truncation error. Derive a maximum principle; show that by suitable choice of the constants C and D the maximum principle may be applied to the mesh function $U_{r,s} - u(x_r, y_s) + Cx_r^2 + D$ to give a bound on the error of the numerical solution.

6.5   The function $u(x,y)$ satisfies the equation $u_{xx} + u_{yy} + f(x,y) = 0$ in the sector of the circle defined by $0 \leq x^2 + y^2 \leq 1$, $0 \leq y \leq x$. A Neumann condition is given on the boundary $y = x$, and Dirichlet conditions on the rest of the boundary. Using a uniform square mesh of size $\Delta x = \Delta y = \frac{1}{3}$ leads to a system of linear equations of the the form $Au = b$; construct explicitly the elements of the matrix A.

Transform the problem into polar co-ordinates, and construct the matrix of a similar system of linear equations.

6.6   Suppose that u is the solution of $u_{xx} + u_{yy} = f$ in the unit square with a Neumann condition $u_x = g$ on $x = 1$ and Dirichlet conditions on the other three sides. A numerical solution is obtained from the usual central difference approximation on a uniform square mesh, with the boundary condition approximated by

$$\frac{U_{J+1,s} - U_{J-1,s}}{2\Delta x} = g_s.$$

Show that

$$U_{r,s} = u(x_r, y_s) + (\Delta x)^2 \psi(x_r, y_s) + O\big((\Delta x)^3\big),$$

where ψ satisfies

$$\psi_{xx} + \psi_{yy} = -\tfrac{1}{12}(u_{xxxx} + u_{yyyy})$$

and give the boundary conditions satisfied by $\psi(x, y)$.

Obtain a similar result for the case where the grid is uniform and rectangular, with $\Delta y = \tfrac{1}{2}\Delta x$.

6.7   The θ-method is used to solve the heat equation $u_t = u_{xx}$, with Dirichlet boundary conditions at $x = 0$ and $x = 1$; suppose that we are given the bound on the truncation error

$$|T_j^{n+1/2}| \le C(1 - \alpha\Delta t)^n, \quad n \ge 0, \;\; 0 < j < J,$$

where α and C are positive constants such that $\alpha\Delta t < 1$ and $\theta\alpha\Delta t < \tfrac{1}{2} - \tfrac{1}{8}\alpha$. Use a comparison function of the form

$$(1 - \alpha\Delta t)^n x_j(1 - x_j)$$

to obtain the error bound

$$|e_j^n| \le K(1 - \alpha\Delta t)^n x_j(1 - x_j),$$

and express the constant K in terms of C.

Repeat the analysis with $\sin \pi x_j$ replacing $x_j(1 - x_j)$.

7

Iterative solution of linear algebraic equations

In Chapter 6 we have discussed alternative methods for approximating the solution of linear elliptic equations, the finite difference method, using either the differential equation or an integral form, and the finite element method. Each of these methods gives rise to a system of linear algebraic equations, which may be very large. A two-dimensional problem may lead to a system of several thousand unknowns, and three-dimensional problems involving several hundred thousand unknowns are common in real engineering situations. The solution of such a system is a major problem in itself and has been the subject of much detailed study. As we have seen above, the system of equations produced by a discretisation has many special features and an efficient solution procedure must exploit these. The most obvious property of the system is that it is extremely sparse. Even when there are many thousand unknowns, each equation will involve one unknown and the unknowns at its immediate neighbours. In particular, if we write the equations in the conventional notation

$$A\mathbf{x} = \mathbf{b}, \tag{7.1}$$

where A is an $N \times N$ matrix, $\mathbf{b}$ the given data vector and $\mathbf{x}$ the vector of N unknown interior mesh values, there is an implied one-dimensional ordering of these values which is somewhat unnatural and obscures the important property that only immediate neighbours are involved. Each row of the matrix A involves only a very small number of non-zero elements, commonly five or seven; moreover in many problems a suitable ordering of the unknowns will lead to a matrix in which these non-zero elements occur in a regular pattern. In devising efficient methods for the solution of the system these structural properties will be important, as well as the properties of the signs of the elements which we have emphasised in Chapter 6.

The most obvious method of solving (7.1) is direct Gaussian elimination. For one-dimensional problems ordered in a natural way this is very efficient: a second order equation approximated by a second order accurate method leads to a tridiagonal system of equations and the very straightforward Thomas algorithm presented in Section 2.9 is equivalent to Gaussian elimination without pivoting; even if a higher order difference scheme were used, so that more than three neighbouring points were involved in each difference equation, the system of equations would be 'banded' rather than tridiagonal (that is $A_{lm} \neq 0$ for only small values of $|l - m|$) and direct elimination in the natural order would be very efficient.

However, in two and three dimensions any ordering will give a very much larger bandwidth. Consider, for example, the solution of Poisson's equation on a rectangular region, covered with a uniform rectangular mesh, so that the unknown interior values are $U_{r,s}$ for $r = 1, 2, \ldots, J_x$ and $s = 1, 2, \ldots, J_y$. We number these points in the standard or *natural ordering*; starting at the bottom, numbering within each row from left to right, and moving up the rows in order. The unknown $U_{r,s}$ will thus be number $k = r + J_x(s-1)$ in this sequence. Then a general equation of the system will involve $U_{r,s}$ and its four neighbours on the mesh, which will be unknowns number k, $k + 1$, $k - 1$, $k + J_x$ and $k - J_x$. The non-zero elements in the matrix therefore lie within a band centred on the diagonal and of width $2J_x + 1$. During the elimination process for the solution of the system, with no pivoting, the elements outside this band will remain zero, but the elements within the band will fill up with non-zero elements. The total number of unknowns is $J_x J_y$, and the solution process will require of order $J_x J_y (2J_x+1)^2$ arithmetical operations. This is of course very much less than would be required if the matrix were full, of order $(J_x J_y)^3$ operations, but is still considerable. On a square mesh of size h the number of arithmetical operations required is of order h^{-4} for a two-dimensional problem, and of order h^{-7} for a three-dimensional problem. In two dimensions, the flexibility of the elimination process in dealing with arbitrary regions and meshes has made it the preferred method for finite element equations, but its dominance here is now being challenged by iterative methods and it is widely recognised that these will be of increasing importance in three-dimensional problems.

When the direct elimination method is applied to a sparse matrix the amount by which the matrix fills up with non-zero elements depends on the ordering of the rows of the matrix. The study of this effect is well developed, and the natural ordering is by no means the best for our model problem. The use of an optimum ordering strategy makes it possible to use direct elimination for quite large problems; the solution of sparse systems by direct methods is described in detail in Duff, Erisman and Reid (1986).

Iterative methods make maximum use of the structure of the matrix A. They underwent very rapid development in the 1950's as part of the effort in the infant nuclear energy industry to develop efficient methods for modelling neutron transport. We will devote the first part of this chapter to giving a somewhat simplified account of these methods and their theory; then we go on to describe in Sections 7.6 and 7.7 much more efficient modern methods, which make some use of these basic iterative processes.

7.1 Basic iterative schemes in explicit form

Suppose we write the familiar five-point scheme on a two-dimensional mesh in the form

$$\tilde{c}_P U_P - [\tilde{c}_E U_E + \tilde{c}_W U_W + \tilde{c}_N U_N + \tilde{c}_S U_S] = b_P \qquad (7.2)$$

where $U_P, U_E, \ldots, U_S$ are unknown (interior) values of U and b_P represents all the known data, including boundary values given by Dirichlet boundary conditions. Thus if, for example, U_N is a known value on the boundary this term would be transferred to the right-hand side of (7.2); it would not appear on the left and $\tilde{c}_N$ would be zero in (7.2). It is because of this difference in convention from that in Chapter 6 that we have used the coefficients $\tilde{c}_P$ etc. here rather than c_P; this distinction corresponds to what happens in the practical implementation of finite element methods, where the matrix elements A_{ij} as in (6.116) are first calculated without reference to the boundary conditions that are to be applied, and then those corresponding to known values U_j are transferred to the right-hand side. The present convention is also closer to that used for evolutionary problems in Chapter 5.

The iteration starts from an initial estimate of all the unknowns, which we denote by $U_P^{(0)}, U_E^{(0)}, \ldots, U_S^{(0)}$. The simplest iterative procedure, first used by Jacobi in 1844 and known as the *Jacobi method* or the *method of simultaneous displacements*, gives the successive iterates $U_P^{(n)}$ defined by

$$U_P^{(n+1)} = (1/\tilde{c}_P)[b_P + \tilde{c}_E U_E^{(n)} + \tilde{c}_W U_W^{(n)} + \tilde{c}_N U_N^{(n)} + \tilde{c}_S U_S^{(n)}],$$
$$n = 0, 1, 2, \ldots . \quad (7.3)$$

Clearly the algorithm is independent of the ordering of the unknowns, and it is admirably suited to modern parallel computers. Also if the coefficients are all positive with $\tilde{c}_P$ at least as great as the sum of the rest, as in Section 6.3, subtracting $(1/\tilde{c}_P)$ of (7.2) from (7.3) gives

$$|U_P^{(n+1)} - U_P| \le \max_{E,W,N,S} \{|U_Q^{(n)} - U_Q|\}. \tag{7.4}$$

This is not quite sufficient to show that the iteration converges, as that would require at least a strict inequality. However it does at least show that the sequence does not diverge, and indicates the importance of this property of the coefficients.

On a serial computer, however, the new values of the unknowns $U_P^{(n+1)}$ are calculated from (7.3) in some particular order and then it would seem advantageous to use the latest available value for $U_E, U_W, \ldots, U_S$: one might expect that this will speed convergence as well as mean that only one value at each mesh point needs to be held in the computer memory at any one time. If we use the natural order mentioned above, then (7.3) is replaced by

$$U_P^{(n+1)} = (1/\tilde{c}_P)[b_P + \tilde{c}_E U_E^{(n)} + \tilde{c}_W U_W^{(n+1)} + \tilde{c}_N U_N^{(n)} + \tilde{c}_S U_S^{(n+1)}],$$
$$n = 0, 1, 2, \ldots. \tag{7.5}$$

Clearly, under the same conditions we will again have (7.4). This iterative method is known as the method of *successive displacements* or the *Gauss–Seidel method* after Gauss, who was reported by Gerling in 1843 to have used it, and Seidel who published it independently in 1874: sometimes however it is called after Liebmann who used it in 1918.

Of course in those early days these two procedures were carried out by hand without the aid of electronic computers. Very often the regular application of (7.3) or (7.5) in successive iterations was varied to increase the convergence rate: the equation residuals were usually monitored and possibly the largest chosen to determine which U_P should be changed next. Such so-called relaxation methods reached a high state of development in the 1940's here at Oxford under the leadership of Southwell. One result was a modification of the Gauss–Seidel procedure which is now called *successive over-relaxation* or the *SOR method*. Taking the new value of U_P to be a weighted average of the old value and the value given by (7.5) we obtain, using weights ω and $(1 - \omega)$,

$$
\begin{aligned}
U_P^{(n+1)} &= (1 - \omega)U_P^{(n)} + (\omega/\tilde{c}_P)[b_P + \tilde{c}_E U_E^{(n)} + \tilde{c}_W U_W^{(n+1)} \\
&\qquad + \tilde{c}_N U_N^{(n)} + \tilde{c}_S U_S^{(n+1)}] \\
&= U_P^{(n)} + (\omega/\tilde{c}_P)[b_P + \tilde{c}_E U_E^{(n)} + \tilde{c}_W U_W^{(n+1)} \\
&\qquad + \tilde{c}_N U_N^{(n)} + \tilde{c}_S U_S^{(n+1)} - \tilde{c}_P U_P^{(n)}] \\
&= U_P^{(n)} - (\omega/\tilde{c}_P)r_P^{(n)},
\end{aligned}
\tag{7.6}
$$

where $r_P^{(n)}$ is the residual at point P; this residual is the difference between the left- and right-hand sides of the equation, calculated using the most recent values of the unknowns. Another way of expressing the same procedure is to calculate the correction which would be given by the Gauss–Seidel iteration (7.5), and then multiply this correction by ω before adding to the previous value. The term *over-relaxation* then implies that $\omega > 1$.

In the last form (7.6) we can also see immediately the similarity with schemes for solving the corresponding parabolic equation: if the Jacobi method had been used the identification with the usual explicit scheme would have been exact. We could identify ω with $4\Delta t/(\Delta x)^2$ and would expect convergence to be best with the largest value of ω consistent with stability. In the Jacobi case this would be $\omega = 1$: but we could expect this to be increased in (7.6) because of the greater implicitness – hence the term over-relaxation corresponding to taking $\omega > 1$; $\omega < 1$ is called under-relaxation and $\omega = 1$ corresponds to the Gauss–Seidel iteration. Notice that in this method the over-relaxation parameter ω is taken to be constant. The same value is used, not only for each unknown, but also for each iteration.

7.2 Matrix form of iteration methods and their convergence

In the notation of (7.1) we write the matrix of the system of equations in the form

$$
A = D - L - U
\tag{7.7}
$$

where D is a diagonal matrix of strictly positive elements (corresponding to the coefficients $\tilde{c}_P$), L is strictly lower triangular and U is strictly upper triangular. The matrices L and U are written with the negative sign because in the problems which arose in Chapter 6 the matrix elements off the diagonal are always negative or zero. The ordering of the unknowns completely determines which coefficients of the difference equations appears in L and which in U. Now we can write the three schemes introduced earlier as:

Jacobi:

$$\mathbf{x}^{(n+1)} = D^{-1}[\mathbf{b} + (L + U)\mathbf{x}^{(n)}] \tag{7.8}$$

Gauss–Seidel:

$$D\mathbf{x}^{(n+1)} = \mathbf{b} + L\mathbf{x}^{(n+1)} + U\mathbf{x}^{(n)}, \tag{7.9a}$$

i.e.,

$$\mathbf{x}^{(n+1)} = (D - L)^{-1}[\mathbf{b} + U\mathbf{x}^{(n)}] \tag{7.9b}$$

or

$$\mathbf{x}^{(n+1)} = (I - D^{-1}L)^{-1}[D^{-1}\mathbf{b} + D^{-1}U\mathbf{x}^{(n)}] \tag{7.9c}$$

SOR:

$$D\mathbf{x}^{(n+1)} = (1 - \omega)D\mathbf{x}^{(n)} + \omega[\mathbf{b} + L\mathbf{x}^{(n+1)} + U\mathbf{x}^{(n)}], \tag{7.10a}$$

i.e.,

$$\mathbf{x}^{(n+1)} = (D - \omega L)^{-1}[\omega\mathbf{b} + \omega U\mathbf{x}^{(n)} + (1 - \omega)D\mathbf{x}^{(n)}] \tag{7.10b}$$

or

$$\mathbf{x}^{(n+1)} = (I - \omega D^{-1}L)^{-1}[\omega D^{-1}\mathbf{b} + \{\omega D^{-1}U + (1 - \omega)I\}\mathbf{x}^{(n)}]. \tag{7.10c}$$

These forms will be useful for subsequent analysis but the earlier explicit forms indicate more clearly how each procedure is actually carried out.

Each of the above iterative procedures can be written

$$\mathbf{x}^{(n+1)} = G\mathbf{x}^{(n)} + \mathbf{c} \tag{7.11}$$

where the matrix G is called the *iteration matrix* for the method. The iteration matrix G can be readily obtained from (7.8), (7.9) or (7.10) and the solution $\mathbf{x}$ of the system satisfies

$$(I - G)\mathbf{x} = \mathbf{c}. \tag{7.12}$$

Then if $\mathbf{e}^{(n)} := \mathbf{x}^{(n)} - \mathbf{x}$ is the error after n iterations,

$$\mathbf{e}^{(n+1)} = G\mathbf{e}^{(n)} = G^2\mathbf{e}^{(n-1)} = \ldots = G^{n+1}\mathbf{e}^{(0)}. \tag{7.13}$$

Lemma 7.1 *The iteration (7.11)–(7.13) converges as $n \to \infty$ for all starting vectors* **x** *if and only if*

$$\rho(G) < 1, \tag{7.14}$$

where $\rho(G)$ is the spectral radius of G, i.e. $\rho(G) = \max_i |\lambda_i(G)|$. The **asymptotic convergence rate** *is $-\ln\rho$.*

Proof For convergence from all data it is necessary and sufficient that $\|G^n\| \to 0$ as $n \to \infty$. Suppose first that the matrix G has a full set of eigenvectors, so there exists a non-singular matrix S such that

$$G = S\Lambda S^{-1} \tag{7.15}$$

where Λ is the diagonal matrix whose diagonal elements are the eigen-values $\{\lambda_i\}$ of G. Then

$$G^n = S\,\Lambda^n\,S^{-1}. \tag{7.16}$$

Now the matrix Λ^n is also a diagonal matrix, with diagonal elements λ_i^n, and these tend to zero if and only if each $|\lambda_i| < 1$.

If G does not have a full set of eigenvectors it must be written in a form similar to (7.15), but with Λ replaced by J, the Jordan form of G. The argument is then essentially similar. □

The rate of convergence, defined by $R = -\ln\rho$, indicates the number of iterations required to obtain a required accuracy. For example, to reduce the error by a factor of 10^p, the number of iterations required will be approximately $(p/R)\ln 10$. Strictly speaking, R is called the asymptotic rate of convergence because it gives an accurate indication of the number of required iterations only for large n; in the early stages of the iteration the error reduction may be faster or slower than that indicated. This follows from (7.13) and (7.15) because

$$\|\mathbf{e}^{(n)}\| \le \|G^n\|\,\|\mathbf{e}^{(0)}\| \le \|S\|\,\|S^{-1}\|\,\|\Lambda\|^n\|\mathbf{e}^{(0)}\|, \tag{7.17a}$$

i.e.,

$$\left(\|\mathbf{e}^{(n)}\|/\|\mathbf{e}^{(0)}\|\right)^{1/n} \le \left(\|S\|\,\|S^{-1}\|\right)^{1/n}\rho; \tag{7.17b}$$

while for the eigenvector corresponding to the largest eigenvalue the error reduction is exactly ρ.

It is now useful to relate more formally the matrix concepts of diagonal dominance and the maximum principle used in the analysis of difference schemes in Chapter 6. A matrix is said to be *diagonally dominant* if, in each row, the diagonal element is at least as large in magnitude as the sum of the moduli of the off-diagonal elements; that is,

$$|A_{ll}| \geq \sum_{m \neq l} |A_{lm}| \quad \forall l, \tag{7.18}$$

which is implied by the condition (6.44) on a difference scheme. It is said to be *strictly diagonally dominant* if strict inequality holds for all l in (7.18). It is then quite easy to show that the Jacobi iteration converges for a strictly diagonally dominant matrix. However, this is not a very helpful result, for our matrices almost never satisfy this condition.

A matrix is said to be *irreducibly diagonally dominant* if (7.18) holds, with strict inequality for *at least one* value of l, and if it is *irreducible*. The concept of irreducibility is closely related to the property of a set of mesh points being connected as in Chapter 6. Rows l and m of the matrix are said to be *connected* if the element A_{lm} is non-zero. The matrix is then *irreducible* if, for any two rows l_1 and l_k, there is a sequence $l_1, l_2, \ldots, l_k$ such that each row of the sequence is connected to the next. Evidently a difference scheme on a set of mesh points which is connected leads to an irreducible matrix; moreover the condition that we must have strict inequality for at least one row corresponds to our earlier condition that a Dirichlet condition must be given on at least part of the boundary. These conditions are thus contained in those required in Chapter 6 for a maximum principle to hold.

Theorem 7.1 *The Jacobi iteration converges for an irreducibly diagonally dominant matrix.*

Proof Suppose that we denote by μ any eigenvalue of the Jacobi iteration matrix, with $\mathbf{v}$ the corresponding eigenvector; then we have to show that $|\mu| < 1$, the strict inequality being essential. We have

$$D^{-1}(L + U)\mathbf{v} = \mu \mathbf{v}, \tag{7.19a}$$

or

$$(\mu D - L - U)\mathbf{v} = \mathbf{0}. \tag{7.19b}$$

We use a method of proof similar to that used in Lemma 6.1 to establish a maximum principle. Thus suppose $|\mu| = 1$ and v_k is an element of the eigenvector $\mathbf{v}$ with the largest magnitude. Then we have

$$\mu A_{kk} v_k = - \sum_{m \neq k} A_{km} v_m, \tag{7.20}$$

from which we deduce

$$|\mu| \leq \sum_{m \neq k} \frac{|A_{km}|}{|A_{kk}|} \frac{|v_m|}{|v_k|}. \qquad (7.21)$$

Thus if $|\mu| = 1$ we must have both equality in the diagonal dominance relation (7.18) for $l = k$, and also $|v_m| = |v_k|$ for all rows m which are connected to k. By repeating the argument with each m replacing k, and then continuing this process recursively, by the irreducibility hypothesis we cover all rows. But inequality must hold for at least one value of l in (7.18); so the assumption that $|\mu| = 1$ leads to a contradiction. $\qquad \square$

We can also give a bound on the values of the relaxation parameter ω in the SOR method.

Theorem 7.2 *The SOR method does not converge unless* $0 < \omega < 2$.

Proof The eigenvalues of the SOR iteration matrix satisfy

$$\det \left[(D - \omega L)^{-1}(\omega U + (1 - \omega)D) - \lambda I\right] = 0 \qquad (7.22a)$$

which may be written

$$\det \left[\lambda(D - \omega L) - \omega U - (1 - \omega)D\right] = 0 \qquad (7.22b)$$

or

$$\det \left[(\lambda + \omega - 1)I - \lambda \omega D^{-1}L - \omega D^{-1}U\right] = 0. \qquad (7.22c)$$

If we expand this determinant we shall obtain a polynomial in λ, and the leading term and constant term can only come from the diagonal elements. The polynomial has the form

$$\lambda^N + \cdots + (\omega - 1)^N = 0, \qquad (7.23)$$

where N is the order of the matrix. Hence the product of the eigenvalues is $(1 - \omega)^N$, so that if $|1 - \omega| \geq 1$ at least one of the eigenvalues must satisfy $|\lambda| \geq 1$, and the iteration will not converge. Note that this result makes no assumptions about the matrix, which can be quite general. $\qquad \square$

It is worth emphasising here that we are using the concept of convergence in this chapter in an entirely different sense from that used in Chapters 2–5, and also that in Chapter 6. In all of those chapters we were concerned with the convergence of U to u as the mesh is refined: at each stage of the convergent process we had a finite-dimensional approximation, but the dimension increased without limit as the mesh was refined. Here we have a fixed mesh and a strictly finite-dimensional algebraic problem which we are solving by an iterative process: only when this has converged do we actually have the approximation U to u on the mesh. Refining the mesh to get U to converge to u introduces further algebraic problems which could also be solved by iteration. Thus we are looking at convergent processes set within a larger scheme of convergent processes. The distinction is important to make because of the similarity of some of the iterative methods to the evolutionary schemes used to solve parabolic problems, as already remarked. This is reflected in the notation $U^{(n)}$ etc.: thus here we have $n \to \infty$ with Δx fixed; in earlier chapters we had $\Delta t, \Delta x \to 0$ and as a consequence $n \to \infty$.

7.3 Fourier analysis of convergence

For particular model problems we can use Fourier analysis to examine more closely the rate of convergence of various iterative methods. Suppose we are solving the five-point difference scheme approximating $\nabla^2 u + f = 0$ on the unit square with Dirichlet boundary conditions. With a square mesh of size Δx where $J \Delta x = 1$, we have $N = (J-1)^2$ unknown interior values. Their errors can be expanded in terms of Fourier modes, just the sine modes in the case of Dirichlet boundary conditions considered here; at the mesh point (x_r, y_s) we have

$$e_{r,s}^{(n)} = \sum_{k_x, k_y} a^{(n)}(k_x, k_y) \sin k_x x_r \sin k_y y_s \qquad (7.24a)$$

where

$$k_x, k_y = \pi, 2\pi, \ldots, (J-1)\pi. \qquad (7.24b)$$

For the Jacobi method of (7.3), in which $\tilde{c}_P = 4$ and $\tilde{c}_E = \tilde{c}_W = \tilde{c}_N = \tilde{c}_S = 1$, it is clear from the trigonometric formula for $\sin A + \sin B$ that we have

$$e_{r,s}^{(n+1)} = \sum_{k_x, k_y} a^{(n)}(k_x, k_y) \tfrac{1}{4}(2 \cos k_x \Delta x + 2 \cos k_y \Delta x) \sin k_x x_r \sin k_y y_s.$$
$$(7.25)$$

Hence $\sin k_x x_r \sin k_y y_s$ is an eigenvector of the Jacobi iteration matrix $G_J := D^{-1}(L + U)$ with the eigenvalue

$$\mu_J(k_x, k_y) = \tfrac{1}{2}(\cos k_x \Delta x + \cos k_y \Delta x). \qquad (7.26)$$

The choices given by (7.24b) give all of the eigenvectors; and hence the largest values of $|\mu_J|$ are obtained from the extreme values $k_x = k_y = \pi$ or $(J - 1)\pi$. These give $\mu_J = \pm \cos \pi \Delta x$, so that

$$\max |\mu_J| = \cos \pi \Delta x \sim 1 - \tfrac{1}{2}(\pi \Delta x)^2 + \cdots . \qquad (7.27)$$

Clearly this implies very slow convergence for small values of Δx, and it is the lowest frequency mode, $\sin j\pi\Delta x$ in each direction, which converges most slowly: this also occurs for the highest frequency which has the form $\sin j(J-1)\pi\Delta x = (-1)^{j-1}\sin j\pi\Delta x$. However in the latter case we have $\mu_J \sim -1$ and such an error mode could be easily damped by averaging two successive iterations.

The rate of convergence is

$$
\begin{aligned}
-\ln |\mu_J| &= -\ln(\cos \pi\Delta x) \\
&\sim -\ln \left(1 - \tfrac{1}{2}(\pi\Delta x)^2\right) \\
&\sim \tfrac{1}{2}(\pi\Delta x)^2 .
\end{aligned}
\qquad (7.28)
$$

With a mesh size of $\Delta x = 0.02$, for example, this gives a value of 0.00197, so that 1166 iterations are required to reduce the error by a factor of 10. If the original estimate had errors of order 1 and we wish to find the solution to six decimal places we should therefore require nearly 7000 iterations. It is therefore important to investigate other methods, which may converge faster.

The situation with the SOR iteration (7.6) is, however, rather more complicated because $\sin k_x x_r \sin k_y y_s$ is not an eigenvector of the iteration matrix. As with the stability analysis, it is more convenient to work with the complex form; so suppose we look for errors of the form

$$e_{r,s}^n = [g(\xi,\eta)]^n e^{i(\xi r + \eta s)} \qquad (7.29)$$

where $\xi = k_x\Delta x, \eta = k_y\Delta y$. Substitution into (7.6) shows that this gives an eigenvector provided that

$$g(\xi,\eta) = \frac{1 - \omega + \tfrac{1}{4}\omega(e^{i\xi} + e^{i\eta})}{1 - \tfrac{1}{4}\omega(e^{-i\xi} + e^{-i\eta})}. \qquad (7.30)$$

We see that $g(\pm\xi,\pm\eta) \neq g(\xi,\eta)$, showing again why $\sin \xi r \sin \eta s$ is not an eigenvector of the iteration matrix.

However, the situation can be rescued by tilting the space–time mesh as originally proposed by Garabedian and followed up subsequently by LeVeque and Trefethen.[1] We introduce a new index and a form for the

---

[1] LeVeque, R.J. and Trefethen, L.N. (1988), Fourier analysis of the SOR iteration, *IMA J. Numer. Anal.* **8**(3), 273–9.

eigenvector

$$\nu = 2n + r + s, \tag{7.31a}$$

$$e_{r,s}^{\nu} = [g(\xi, \eta)]^{\nu} e^{i(\xi r + \eta s)}. \tag{7.31b}$$

In terms of the new index the error equation corresponding to (7.6) involves three levels, taking the form

$$e_{r,s}^{\nu+2} = (1 - \omega)e_{r,s}^{\nu} + \tfrac{1}{4}\omega[e_{r+1,s}^{\nu+1} + e_{r-1,s}^{\nu+1} + e_{r,s-1}^{\nu+1} + e_{r,s+1}^{\nu+1}]. \tag{7.32}$$

Thus substitution of (7.31b) gives a quadratic for $g(\xi, \eta)$,

$$g^2 = (1 - \omega) + \tfrac{1}{2}\omega[\cos\xi + \cos\eta]g, \tag{7.33}$$

in which the coefficients are even functions of ξ and η. The eigenvectors of the iteration matrix are therefore of the form

$$[g(\xi, \eta)]^{r+s} \sin\xi r \sin\eta s \tag{7.34}$$

and the largest eigenvalue is

$$\max|\lambda_{\text{SOR}}| = \max_{\xi, \eta} \max\{|g_+(\xi, \eta)|^2, |g_-(\xi, \eta)|^2\} \tag{7.35}$$

where g_+ and g_- are the roots of (7.33).

It is now clear that this result could have been obtained directly by examining the behaviour of errors of the form

$$\lambda^n [g(\xi, \eta)]^{r+s} \sin\xi r \sin\eta s. \tag{7.36}$$

Substituting into (7.6) shows that we require

$$\lambda \sin\xi r \sin\eta s = 1 - \omega + \tfrac{1}{4}\omega[g \sin\xi(r + 1) \sin\eta s + (\lambda/g) \sin\xi(r - 1) \sin\eta s$$
$$+ g \sin\xi r \sin\eta(s + 1) + (\lambda/g) \sin\xi r \sin\eta(s - 1)]. \tag{7.37}$$

We then obtain the correct trigonometric sums provided that $g = \lambda/g$, and in that case we obtain an eigenvector with eigenvalue given by

$$\lambda = 1 - \omega + \tfrac{1}{2}\omega(\cos\xi + \cos\eta)\lambda^{1/2} \tag{7.38}$$

in agreement with (7.33).

The Gauss–Seidel method, in which $\omega = 1$, is a special case; for then one of the roots of (7.33) is $g = 0$, so there is a multiple zero eigenvalue and the matrix does not have a full set of eigenvectors. However we can take the limit of (7.38) as $\omega \to 1$ and deduce that the eigenvalues of the Gauss–Seidel iteration matrix are zero and the squares of the eigenvalues of the Jacobi iteration matrix. This shows that the Gauss–Seidel method converges exactly twice as fast as the Jacobi method.

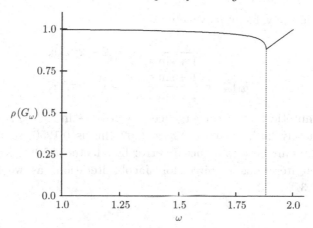

Fig. 7.1. The spectral radius $\rho(G_\omega)$ as a function of ω for $\mu_0 = 0.99$.

More generally, for $\omega \neq 1$, it is easy to confirm from (7.33) that $0 < \omega < 2$ is necessary for $|g_\pm| < 1$. Furthermore, by considering the dependence of $|g_\pm|$ on $\mu = \frac{1}{2}(\cos \xi + \cos \eta)$, it is readily checked that the maximum value is obtained with $\mu_0 = \cos \pi \Delta x$; and then from how this maximum varies with ω we can find the optimal SOR parameter.

The eigenvalues of the SOR matrix satisfy (7.38). For a fixed value of ω the largest eigenvalue corresponds to $\frac{1}{2}(\cos \xi + \cos \eta) = \mu_0 = \cos \pi \Delta x$. Equation (7.38) can then be written

$$\lambda^2 + 2\lambda(\omega - 1 - \tfrac{1}{2}\mu_0^2\omega^2) + (\omega - 1)^2 = 0. \tag{7.39}$$

It is clear that when $\mu_0^2\omega^2 \leq 4(\omega - 1)$, which can only occur for $\omega > 1$, they form a complex conjugate pair with absolute values equal to $\omega - 1$; otherwise, we see from (7.39) that the sum of the roots is positive and therefore they are both positive. We now easily obtain

$$\lambda_{\text{SOR}} = 1 - \omega + \tfrac{1}{2}\mu_0^2\omega^2$$
$$+ \mu_0\omega(1 - \omega + \tfrac{1}{4}\mu_0^2\omega^2)^{1/2} \text{ if } \omega < 1 + \tfrac{1}{4}\mu_0^2\omega^2, \tag{7.40a}$$
$$\lambda_{\text{SOR}} = \omega - 1 \text{ if } \omega \geq 1 + \tfrac{1}{4}\mu_0^2\omega^2. \tag{7.40b}$$

This behaviour is plotted in Fig. 7.1 for a typical value of μ_0. Since the expression (7.40a) decreases with ω and that in (7.40b) increases, the optimum value occurs when $\omega = 1 + \tfrac{1}{4}\mu_0^2\omega^2$, giving

$$\omega = \frac{2}{\mu_0^2}\left(1 - \sqrt{(1 - \mu_0^2)}\right). \tag{7.41}$$

Inserting the value of μ_0 we obtain

$$\omega_{opt} = \frac{2}{1 + \sin \pi \Delta x} \sim 2 - 2\pi \Delta x,$$

$$\rho(G_\omega)_{opt} = \frac{1 - \sin \pi \Delta x}{1 + \sin \pi \Delta x} \sim 1 - 2\pi \Delta x. \qquad (7.42)$$

The asymptotic rate of convergence, given by $- \ln \rho(G_\omega)$, is therefore approximately $2\pi \Delta x$. With $\Delta x = 0.02$ this is 0.1256, so that only 18 iterations are need to reduce an error by a factor of 10; this compares with 1166 iterations required for Jacobi iteration, as we found in Section 7.3.

7.4 Application to an example

The use of the optimum relaxation parameter gives a very significant improvement in the rate of convergence. In our model problem we could calculate the eigenvalues of the Jacobi matrix, and thus find the optimum parameter. In a more general problem this would not be possible, and it is necessary to make some sensible choice of ω. In Fig. 7.2 we show the rate of convergence, for the model problem, as a function of ω over the range $1 < \omega < 2$. It is clear from this fairly typical example that the optimum is quite sharply defined, and any deviation from it unfortunately leads to a substantial reduction of the rate of convergence.

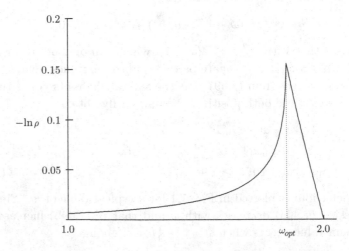

Fig. 7.2. Rate of convergence of the SOR iteration as a function of ω.

However, it does also show that it is better to overestimate than to underestimate the parameter. When $\omega > \omega_{opt}$ the eigenvalue is complex and has modulus $\omega - 1$; the slope of the curve in Fig. 7.2 is close to -1. But on the left of the critical value the slope tends to infinity, and a small deviation in ω has a greater effect.

As an example of the behaviour of the SOR iteration we apply the method to the solution of the linear equations arising from the problem discussed at the end of Chapter 6. Here we used the standard five-point difference scheme to approximate Laplace's equation on a uniform square mesh. The Fourier analysis of Section 7.3 does not strictly apply, since the eigenvector in (7.24a) does not satisfy the correct conditions on the curved boundary. We might expect the effect of the boundary to be fairly small, and therefore that the choice (7.42) for the parameter would not be exactly optimal, but would give good convergence.

This calculation used the interval $\Delta x = 0.025$, and the radius of the circular boundary is 0.4. The iterations started from an initial vector which is zero at all internal points. Fig. 7.3 shows the behaviour of the error as the iteration proceeds, for various choices of the relaxation parameter. To determine the error at each stage we have first carried out a large number of iterations so that the sequence has converged to the true solution $\{U_{r,s}\}$ of the algebraic system. After each iteration the error is then measured by $E^{(n)}$, where

$$E^{(n)} = \left[\frac{1}{N} \sum_{r,s} (U_{r,s}^{(n)} - U_{r,s})^2 \right]^{1/2}. \tag{7.43}$$

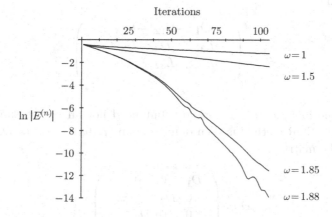

Fig. 7.3. Convergence of SOR iteration.

The graphs show $E^{(n)}$ as a function of n. The graph shows very clearly the great improvement in convergence over Gauss–Seidel iteration, with $\omega = 1$. The optimum choice of parameter predicted by (7.42), with $\Delta x = 0.025$, is 1.8545, and this gives good convergence. However, we find that the slightly different value $\omega = 1.88$ gives a significant improvement.

Our use of the spectral radius of the iteration matrix as a measure of the rate of convergence does not always give an exact picture of what actually happens. If we expand the error in terms of the eigenvectors of the matrix, as in (7.24a), each contribution is multiplied by the corresponding eigenvalue at each iteration. If the largest eigenvalue is significantly larger than all the others, its contribution will soon come to dominate the behaviour of the error. But in practical problems it is much more likely that there will be a number of eigenvalues with nearly the same modulus, and all of them will be making a significant contribution to the error. With a choice of ω close to the optimum, these eigenvalues are likely to be complex, leading to the sort of erratic behaviour shown in Fig. 7.3. For the smaller values of ω most of the eigenvalues are real, and the error behaves more smoothly.

7.5 Extensions and related iterative methods

The convergence of the methods treated in the preceding sections can often be improved, at the expense of some additional computation, by partitioning the matrix A into blocks . Suppose that A is partitioned in the form

$$
A = \begin{pmatrix} D_1 & -U_{12} & -U_{13} & \cdots \\ -L_{21} & D_2 & -U_{23} & \cdots \\ -L_{31} & -L_{32} & D_3 & \cdots \\ & \cdots & \cdots & \cdots \cdots \end{pmatrix}, \tag{7.44}
$$

where the matrices D_k are square, but need not all be the same size. The block SOR method is then defined as in (7.10a), but now D, L and U are the matrices

$$
D = \begin{pmatrix} D_1 & 0 & 0 & \cdots \\ 0 & D_2 & 0 & \cdots \\ 0 & 0 & D_3 & \cdots \\ \cdots & \cdots & \cdots & \cdots \end{pmatrix}, \tag{7.45}
$$

$$L = \begin{pmatrix} 0 & 0 & 0 & \dots \\ L_{21} & 0 & 0 & \dots \\ L_{31} & L_{32} & 0 & \dots \\ \dots & \dots & \dots & \dots \end{pmatrix}, \tag{7.46}$$

$$U = \begin{pmatrix} 0 & U_{12} & U_{13} & \dots \\ 0 & 0 & U_{23} & \dots \\ 0 & 0 & 0 & \dots \\ \dots & \dots & \dots & \dots \end{pmatrix}. \tag{7.47}$$

The practical difference from the point iterative methods discussed previously is that now the calculation of $\mathbf{x}^{n+1}$ from (7.10a) requires the solution of a set of separate linear systems with matrices D_1, D_2, D_3, ... In our model problem the matrix A is of order $(J-1)^2$; numbering the unknowns in the natural order we can partition A into blocks where each D_k is a $(J-1) \times (J-1)$ matrix. All the L_{lm} and U_{lm} matrices are diagonal if $|l - m| = 1$ and otherwise are zero. The diagonal block matrices D_k are tridiagonal, so the calculation of $\mathbf{x}^{n+1}$ requires the solution of $J - 1$ systems of linear equations, each tridiagonal of order $J - 1$; each of these systems can be efficiently solved by the Thomas algorithm.

We might also number the unknowns within each column from bottom to top, moving along the columns in order. Partitioning the new matrix into blocks of the same size gives a new block iteration, treating the unknowns in vertical blocks on the grid instead of horizontal blocks. If we use a combined iterative method, with one horizontal line iteration followed by one vertical line iteration, the result is similar to one time step of the Peaceman–Rachford ADI method of Section 3.2 for the heat equation $u_t = \nabla^2 u$. These become identical if we take $\omega = 1$ in (7.10a), $\nu_x = \nu_y = 1$ in (3.15), and replace $L\mathbf{x}^{(n+1)}$ by $L\mathbf{x}^{(n)}$ in (7.10a); this last modification would replace a method based on Gauss–Seidel iteration by one based on Jacobi iteration.

This method of alternating line iteration is closely related to the ADI method for solving the time-dependent heat equation, and looking for a steady solution for large time. But the iterative method is more general, since we do not need the successive stages of the iteration to agree accurately with solutions of the heat equation at specific times; only the limit solution is required.

However, in many practical application areas the methods described so far have largely been superseded by more efficient and usually more elaborate methods. We end this chapter with a brief introduction to the *multigrid method* and the *conjugate gradient method*; these methods and their derivatives are now most often used in practical computations.

7.6 The multigrid method

This method, originally proposed by Brandt[1] has been developed over the last 20 years into a very efficient method for the solution of systems of algebraic equations derived from partial differential equations. In most cases it gains this efficiency by making use of the underlying differential equation. The fundamental idea is best introduced by means of an example; we shall use an example involving the self-adjoint equation

$$(a(x,y)u_x)_x + (a(x,y)u_y)_y + f(x,y) = 0, \quad (x,y) \in \Omega, \qquad (7.48a)$$

$$u = 0, \quad (x,y) \in \partial\Omega, \qquad (7.48b)$$

where Ω is the unit square

$$\Omega := (0,1) \times (0,1) \qquad (7.49)$$

and $\partial\Omega$ is the boundary of the square. The function a is

$$a(x,y) = x + 2y^2 + 1 \qquad (7.50)$$

and f is chosen so that the solution of the difference equations gives an approximation to

$$u(x,y) = xy(1-x)(1-y) \qquad (7.51)$$

at the mesh points.

We use the central difference approximation (6.5), with $\Delta x = 1/10$, and then with $\Delta x = 1/20$, and first solve each of the resulting system of equations by Gauss–Seidel iteration. In order to bring out the basis of the multigrid method, we start the iteration with the initial vector whose elements are independent random numbers, uniformly distributed in $[0,1]$, though in practice this would not be a useful starting vector. The progress of the iterations is illustrated in Fig. 7.4. The graphs show $\ln(R_n)$, where R_n is the 2–norm of the residual vector after n iterations; the solid line corresponds to $\Delta x = 1/10$, and the dotted line to $\Delta x = 1/20$.

These graphs show two important features: (i) the first few iterations give a rapid reduction in the residual, but then the rate of convergence soon becomes very much slower, and (ii) the eventual rate of convergence

---

[1] Brandt, A. (1977) Multi-level adaptive solutions to boundary value problems. *Math. Comput.* **31**, 333-90.

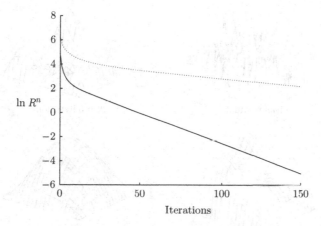

Fig. 7.4. Gauss–Seidel iteration for the model problem: lower curve, $\Delta x = 1/10$; upper curve, $\Delta x = 1/20$.

when $\Delta x = 1/20$ is much slower than when $\Delta x = 1/10$. This second observation is just as predicted by (7.38). We will return to this problem later, in Section 7.8.

Figure 7.5 shows surface plots of the initial residual, and of the residual after 5, 20 and 50 iterations. These plots are not to the same scale, but are arranged to show roughly the same maximum size. It is the shape of the plots which is important, not their size; after the very jagged initial plot they rapidly become smoother, and even after five iterations the random peaks have largely disappeared.

We can now apply these observations to suggest a simple form of the multigrid method. We wish to solve the systems of equations

$$Ax = b. \tag{7.52}$$

Starting from the first estimate $\mathbf{x}^{(0)}$ we construct the sequence $\mathbf{x}^{(k)}$, $k = 1, 2, \ldots$ and the corresponding *residuals*

$$\mathbf{r}^{(k)} = \mathbf{b} - A\mathbf{x}^{(k)}. \tag{7.53}$$

The corresponding *error* $e^{(k)}$ is then the vector $e^{(k)} = \mathbf{x} - \mathbf{x}^{(k)}$, and

$$Ae^{(k)} = \mathbf{r}^{(k)}. \tag{7.54}$$

Now suppose that the residual $\mathbf{r}^{(k)}$ is smooth, in a sense which can be quite precisely defined. Then the elements of the vector $\mathbf{r}^{(k)}$ can be regarded as the values of a smooth function $r^k(x, y)$ at the mesh points (x_r, y_s). It is important here that the rows of the matrix A are correctly

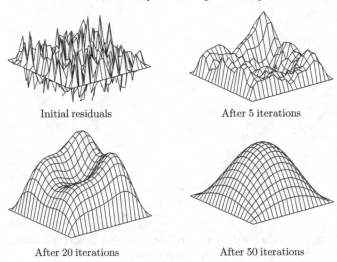

Fig. 7.5. Surface plot of residuals.

scaled, so that the system of algebraic equations represent a difference approximation to the differential equation; the elements of A will contain the factor $1/(\Delta x)^2$. The algebraic equation (7.54) can then be regarded as the finite difference approximation to the differential equation

$$e_{xx} + e_{yy} = r. \qquad (7.55)$$

Assuming that the functions e and r are smooth, we can therefore find an approximate solution of this differential equation on a coarser grid; this coarser grid is often, though not always, obtained by omitting alternate mesh lines in both x and y directions, giving a mesh with twice the mesh size Δx and Δy. The important point is that this finite difference problem now involves a number of unknowns which is a quarter of the number in the original problem. As our analysis above has shown the usual iterative methods will now converge much faster, and each iteration will require much less work.

We have thus outlined the key part of a two-grid method, which consists of four stages.

Smoothing First perform a small number of iterations of an iterative method to smooth the residuals.

Restriction Compute the residuals and transfer them to the coarser grid; this can be done just by using the values at the even-numbered mesh points, or by some weighted mean of neighbouring residuals.

Coarse grid correction Solve the resulting system of equations on the coarser grid, the restricted residuals being the vector on the right hand side.

Prolongation This solution is defined on the coarser grid, and is extended to the fine grid, usually by linear halfway interpolation in both the x- and y-directions. The result is our approximation to the vector $e^{(k)}$, and is therefore added to the fine grid solution x^k to give a better approximation to the solution of our problem.

Post–smoothing It is usual to perform more smoothing iterations at this stage.

The central step in this two-grid method is the solution of the coarse-grid system of equations. The method now becomes a multigrid method when we make it recursive, so that we solve the coarse grid equations in the same way as the fine grid equations, by defining a new level of coarsening of the coarse grid. The recursion continues, each grid having a mesh size double that of the previous grid, until a grid is reached where the number of unknowns is so small that a direct solution of the algebraic equations is as efficient as an iterative method.

In each of the four stages of the multigrid method there is a wide choice; for example, any of the iterative methods described in Section 7.1 might be considered for the smoothing stage, and there are many other possibilities. As a very simple example, consider the application of the weighted Jacobi method to the solution of Poisson's equation in a square region.

The weighted Jacobi method extends the Jacobi method just as SOR extends Gauss–Seidel iteration. It replaces (7.8) by

$$D\mathbf{x}^{(n+1)} = (1 - \omega)D\mathbf{x}^{(n)} + \omega(\mathbf{b} + (L + U)\mathbf{x}^{(n)}).$$

where ω is a factor to be chosen. The Fourier analysis of Section 7.3 determines the eigenvalues of the iteration matrix as in (7.26), but now

$$\mu = 1 - \omega(\sin^2 k_x \Delta x + \sin^2 k_y \Delta y).$$

The smooth eigenvalues will be dealt with by the coarse grid correction, and the smoothing process only needs to consider the Fourier modes with $k_x \geq \frac{1}{2}J\pi$ or $k_y \geq \frac{1}{2}J\pi$, or both. We leave it as an exercise to show that the best choice is now $\omega = \frac{4}{5}$, and that this leads to a reduction of all relevant Fourier modes by a factor better than $\frac{3}{5}$. The important result is that this smoothing factor is independent of the mesh size.

We can now apply the multigrid method to the solution of a particular model problem used in Briggs, Henson and McCormick (2000); this is Poisson's equation on the unit square, with the function f chosen so that the solution is $u(x, y) = (x^2 - x^4)(y^4 - y^2)$, which vanishes on the boundary of the square. We use 2 sweeps of the weighted Jacobi method as smoother, and another two sweeps for post-smoothing; for restriction we use full weighting, calculating the coarse-grid residual as a weighted sum of the nine neighbouring fine-grid residuals. The coarse-grid correction uses one iteration of the same multigrid method; taking $n = 2^p$ intervals in each direction on the finest grid, the coarsest grid has just one internal point, so the solution on this grid is trivial. The initial approximation is taken to be zero everywhere. Fig. 7.6 displays the results of a numerical calculation; it gives $\ln(\|r^{(k)}\|)$ after each iteration. There are five curves for, respectively, $\Delta x = 2^{-p}, p = 4, 5, 6, 7, 8$; the curves are closely parallel, showing that the rate of convergence of the multigrid method is independent of the mesh size (compare with Fig. 7.4). In fact, apart from a slightly faster reduction in the residual after the first two iterations, for the smaller values of p, the curves would be identical.

The multigrid method requires the construction of a coarse grid from a given fine grid. In our example this is straightforward, involving just the even numbered grid lines. If we cover a nonrectangular region with a mesh of triangles, as in Fig. 6.5, it is more convenient to begin by defining the coarsest mesh and then construct a sequence of finer meshes by subdividing each triangle, placing vertices at the midpoints of the sides.

In the smoothing stage of the process we may choose any of a wide range of iterative schemes; the choice will depend on the properties of the problem. For example, when solving a convection-dominated problem such as (6.24), with b and c much larger than a, and probably taking different values, and different signs, in different parts of the region, the simple weighted Jacobi smoothing scheme will not be the most efficient. There are a great many studies reported in the multigrid literature which help in making this choice.

Having chosen a smoothing scheme it is also necessary to decide how many iterations of the smoother should be used – commonly just one or two, but sometimes more. A similar decision has to be made when making the coarse grid correction, applying the multigrid method recursively on the coarse grid; this can be done by using more than one iteration

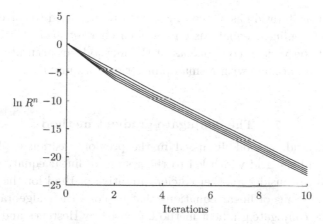

Fig. 7.6. Multigrid convergence for different grid sizes: starting from the lower curve, $\Delta x = 2^{-p}$ for p = 4,5,6,7 and 8.

of the multigrid correction here. Use of one iteration leads to what are known as *V–cycles*, while two iterations lead to *W–cycles*. More than two iterations here are rarely used.

The crucially important property of the multigrid method is that its overall rate of convergence is independent of the grid size. As we illustrated in Fig. 7.4 the classical iterative methods converge more slowly as the grid size is reduced. Theoretical analysis of the multigrid method introduces the idea of a *work unit*, which is the amount of computational work involved in carrying out one relaxation sweep on the finest grid. This work unit is clearly proportional to the number of unknown values at the grid points. The multigrid method should obtain an accurate solution of the problem with just a small number of work units, usually about 10. This aim is achieved over a very wide range of problems.

So far in this chapter we have only discussed linear problems, but of course most practical problems involve nonlinear differential equations. The multigrid method has been extended to solve such problems with the same sort of efficiency. A straightforward approach would be to use Newton's method to solve the full set of nonlinear algebraic equations. Each iteration of Newton's method then involves a large set of *linear* equations, which can be solved by the multigrid method which we have described; this now involves a double iteration. The method more commonly used is the *full approximation scheme*, which is fully described in the books cited at the end of this Chapter. This is a modification of the multigrid scheme as we have just described it. It requires some form of nonlinear iteration as a smoother, and will also need to solve a small system of nonlinear equations on the final coarsest grid. With careful choice of the various components of the method, an accurate solution can still be obtained with a small number of work units.

7.7 The conjugate gradient method

The multigrid method discussed in the previous section exploits the structure of the grid which led to the system of linear equations. Now we go on to consider another efficient iterative method for the solution of large systems of linear equations that exploits their algebraic structure. The conjugate gradient method, devised by Hestenes and Stiefel,[1] applies to systems arising from a larger class of problems which have a symmetric positive definite structure.

We begin with the solution of the system $A\mathbf{x} = \mathbf{b}$, where the matrix A is symmetric and positive definite; in practical problems it will of course also be sparse. The method is based on the observation that the unique minimum of the function

$$\phi(\mathbf{x}) = \tfrac{1}{2}\mathbf{x}^T A\mathbf{x} - \mathbf{b}^T\mathbf{x} \qquad (7.56)$$

is the solution of the system $A\mathbf{x} = \mathbf{b}$; the corresponding value of $\phi(\mathbf{x})$ is zero. The conjugate gradient method proceeds by constructing a sequence of vectors $\mathbf{x}^{(k)}$ for which the values $\phi(\mathbf{x}^{(k)})$ form a monotonic decreasing sequence, converging to zero.

Suppose that we have an estimate $\mathbf{x}^{(k)}$ of the solution, and have some means of choosing a *search direction* defined by the vector $\mathbf{p}^{(k)}$. Then we can choose the next vector in the sequence by

$$\mathbf{x}^{(k+1)} = \mathbf{x}^{(k)} + \alpha_k \mathbf{p}^{(k)}, \qquad (7.57)$$

where the scalar α_k is chosen to minimise $\phi(\mathbf{x}^{(k)})$ as a function of α_k. It is easy to show by differentiation that the required minimum is obtained when

$$\alpha_k = \mathbf{p}^{(k)T}\mathbf{r}^{(k)}/\mathbf{p}^{(k)T}A\mathbf{p}^{(k)}, \qquad (7.58)$$

where $\mathbf{r}^{(k)}$ is the residual vector

$$\mathbf{r}^{(k)} = \mathbf{b} - A\mathbf{x}^{(k)}. \qquad (7.59)$$

---

[1] Hestenes, M.R. and Stiefel, E. (1952), Methods of conjugate gradients for solving linear systems, *J. Res. Nat. Bur. Stand.* **49**, 409–36.

The best choice of the search direction $\mathbf{p}^{(k)}$ is obviously important. The key property of the conjugate gradient method is that it constructs a sequence of vectors $\mathbf{p}^{(k)}$ such that when we determine $\mathbf{x}^{(k+1)}$ from (7.57), with α_k from (7.58), the new vector $\mathbf{x}^{(k+1)}$ gives the minimum of $\phi(\mathbf{x})$ not only in this search direction, but also in the whole subspace spanned by the search directions $\mathbf{p}^{(0)} \ldots \mathbf{p}^{(k)}$. We shall see below that this key property results from the directions being 'A-conjugate', that is that $\mathbf{p}^{(k)T} A \mathbf{p}^{(j)} = 0$ for $j < k$.

This method in its basic form converges, but for the type of problems which we are considering the rate of convergence is usually rather disappointing. The situation is transformed by the use of *preconditioning*. This idea is based on the observation that the solution of $A\mathbf{x} = \mathbf{b}$ is the same as the solution of $PA\mathbf{x} = P\mathbf{b}$ when P is any nonsingular matrix; a suitable choice of P may make the new system of equations much easier to solve. The matrix PA will not usually be symmetric, so we take the idea a step further by choosing a nonsingular matrix C with further properties that we shall discuss below, and writing the system as

$$\left(C^{-1}A(C^{-1})^T\right)C^T\mathbf{x} = C^{-1}\mathbf{b}. \tag{7.60}$$

The matrix of this system is symmetric; having solved this system for the vector $C^T\mathbf{x}$ we need to recover the required vector $\mathbf{x}$. However, the process can be simplified by writing $M = C^TC$, which is now the preconditioner; then the preconditioned conjugate gradient algorithm takes the following form:

starting from $\mathbf{x}^{(0)} = 0$, $\mathbf{r}^{(0)} = \mathbf{b}$, solve $M\mathbf{z}^{(0)} = \mathbf{b}$, and set $\beta_1 = 0, \mathbf{p}^{(1)} = \mathbf{z}^{(0)}$. Then for $k = 1, 2, \ldots$ carry out the following steps:

$$\alpha_k = \mathbf{z}^{(k-1)T}\mathbf{r}^{(k-1)}/\mathbf{p}^{(k)T}A\mathbf{p}^{(k)}$$
$$\mathbf{x}^{(k)} = \mathbf{x}^{(k-1)} + \alpha_k\mathbf{p}^{(k)}$$
$$\mathbf{r}^{(k)} = \mathbf{r}^{(k-1)} - \alpha_k A\mathbf{p}^{(k)} \text{ and solve } M\mathbf{z}^{(k)} = \mathbf{r}^{(k)}$$
$$\beta_{k+1} = \mathbf{z}^{(k)T}\mathbf{r}^{(k)}/\mathbf{z}^{(k-1)T}\mathbf{r}^{(k-1)}$$
$$\mathbf{p}^{(k+1)} = \mathbf{z}^{(k)} + \beta_{k+1}\mathbf{p}^{(k)}. \tag{7.61}$$

Notice that the matrix $C^{-1}A(C^{-1})^T$ is not used in the calculation, which only uses the matrix M and the original matrix A. It can be shown that the residual vectors $\mathbf{r}$ and the search directions $\mathbf{p}$ satisfy the relations

$$\mathbf{r}^{(j)T}M^{-1}\mathbf{r}^{(i)} = 0,$$
$$\mathbf{p}^{(j)T}\left(C^{-1}A(C^{-1})^T\right)\mathbf{p}^{(i)} = 0, \tag{7.62}$$

for $i \neq j$, and these form the basis for the convergence of the process.

The choice of preconditioner is very important for the success of the method. There are two obvious extremes: taking $M = I$, the unit matrix, corresponds to the basic method without preconditioning, which converges too slowly; on the other hand, if we choose $M = A$ it is easy to see that the process terminates with the exact solution after one step. But this is not a practical choice, as we must be able to solve easily the system $M\mathbf{z} = \mathbf{r}$. For a practical method we need a matrix M which is in some sense close to A, but such that $\mathbf{z}$ can be easily calculated.

A frequently used method constructs M by the incomplete Cholesky process, which leads to what is known as the ICCG algorithm. The Cholesky algorithm constructs a lower triangular matrix L such that $A = LL^T$. As with other factorisation methods L is usually a rather full matrix even when A is very sparse. The incomplete Cholesky algorithm uses the same formulae for the calculation of the elements of L_{ij} in the positions where A_{ij} is nonzero, but puts $L_{ij} = 0$ wherever $A_{ij} = 0$. The result is that L has the same sparsity structure as A, and LL^T is fairly close to A. As in other factorisation processes, it is not necessary to construct the preconditioning matrix M itself. Having constructed L, the vector $\mathbf{z}^{(k)}$ can be found by solving two triangular systems,

$$L\eta = \mathbf{r}^{(k)}, \quad L^T\mathbf{z}^{(k)} = \eta. \tag{7.63}$$

The whole calculation is very efficient, because of the sparsity of L.

We have just seen that in the ICCG method it was not necessary to construct the matrix M explicitly, and the same is true more generally. All that we need is to be able to find a vector $\mathbf{z}$ which is close to the solution of $A\mathbf{z} = \mathbf{r}$, and which exactly satisfies $M\mathbf{z} = \mathbf{r}$ where M is some symmetric and positive definite matrix. This means that we might consider any of a number of iterative methods; for instance we might use several iterations of the Jacobi method for solving $A\mathbf{z} = \mathbf{r}$, since the iteration matrix in this case is symmetric. The iteration matrix for the SOR method is not symmetric, though it is still possible that some symmetric matrix M might exist for which the result is the exact solution. In practical problems the efficiency of various preconditioners has to be assessed by experimentation, and in many cases it is found that SOR iteration is quite a good preconditioner. The incomplete Cholesky algorithm seems to be efficient for a wide range of problems, but it makes no use of the special properties of the PDE from which the system of equations was derived. A wide range of possibilities have been studied and recommended, for example using various block methods discussed in Section 7.5, and the related ideas of domain decomposition. The rate of convergence depends crucially on the choice of preconditioner, and it is difficult to give a simple Fourier analysis like that of the multigrid method.

The conjugate gradient method as we have presented it requires that the matrix A is symmetric and positive definite. Extensions of the idea to matrices which are not symmetric and/or not positive definite have been devised, but so far practical experience of them is less satisfactory than for the original version; research in this area is currently very active. An alternative but related approach sets out directly to minimise the sum of squares of the residuals, and can apply to an arbitrary matrix A: the most widely used algorithm is GMRES, due to Saad and Schultz.[1] Like the conjugate gradient algorithm, it constructs a sequence of search directions $\mathbf{p}^{(k)}$, but requires all the previous vectors $\mathbf{p}^{(1)}, \ldots, \mathbf{p}^{(k)}$ for the construction of $\mathbf{p}^{(k+1)}$. Note that the conjugate gradient algorithm constructs the vector $\mathbf{p}^{(k+1)}$ from only $\mathbf{p}^{(k)}$ and $\mathbf{p}^{(k-1)}$, saving in computational labour, and storage requirements. Both methods are members of the general class of *Krylov subspace methods* as they utilise successive powers $A\mathbf{z}^{(0)}, A^2\mathbf{z}^{(0)}, A^3\mathbf{z}^{(0)}, \ldots$ of an initial vector.

7.8 A numerical example: comparisons

To conclude this chapter, we show in Fig. 7.7 the result of applying three iterative methods to the problem used in Fig. 7.4:

$$(a(x,y)u_x)_x + (a(x,y)u_y)_y + f(x,y) = 0, \quad 0 < x < 1, \ 0 < y < 1,$$
$$(7.64)$$

where $a(x,y) = (x + 2y^2 + 1)$, with solution $u(x,y) = x(1-x)y(1-y)$. The numerical experiments used $J = 40$, so the system of algebraic equations is of order 1521.

First we use the symmetric SOR method described in Section 7.2. Since the coefficient $a(x,y)$ is not a constant the problem is more general than that in Section 7.6 and we cannot use Fourier analysis to find the optimal relaxation parameter ω. We experimented with a range of values, and found empirically that $\omega = 1.85$ gave the fastest convergence; this is in fact close to the optimum value which would be predicted by (7.42). In Fig. 7.7 the top graph shows the convergence of the iteration;

[1] Saad, Y. and Schultz, M.H.(1986) A generalised minimal residual algorithm for solving nonsymmetric linear systems. *J. Sci. Stat. Comput.* **7**, 856–69.

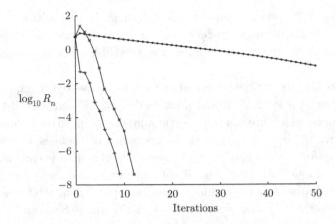

Fig. 7.7. Comparison of iterations for the model problem (7.64).
Top curve: SOR with $\omega = 1.85$.
Middle curve: Incomplete Cholesky Conjugate Gradient.
Bottom curve: Multigrid.

as expected it is slow, and after 50 iterations the residual has been reduced by a factor of about 55.

The bottom curve shows the result from use of a multigrid method, with Gauss–Seidel iteration as a smoother and two stages of coarsening. This gives a solution correct to 6 decimal places after only 8 iterations. The middle curve uses the preconditioned conjugate gradient method, with incomplete Cholesky factorisation as the preconditioner: after the first iteration this indicates a rate of convergence very similar to that of multigrid.

It is not the purpose of this calculation to compare multigrid with ICCG, but to show that both of them are far more efficient than the classical SOR iteration. Of course, each of them is also more complicated, so that one iteration requires more work than one SOR iteration. Nevertheless even allowing for this, both of them are orders of magnitude faster than SOR.

The two methods, multigrid and conjugate gradient should be regarded as complementary rather than as in competition. Recent studies reported in Elman, Silvester and Wathen (2004) of the solution of the Navier–Stokes equations for steady incompressible fluid flow use the conjugate gradient method, with multigrid iterations as the preconditioner. The results demonstrate convergence within a small number of work units with very little dependence on the number of mesh points.

Bibliographic notes and recommended reading

A full treatment of the classical theory of iterative methods for linear equations is given by Varga (1974), and also in the slightly earlier book by Young (1971). Direct methods that make full use of matrix sparsity are fully treated in Duff, Erisman and Reid (1986). Golub and Van Loan (1996) and Hackbusch (1994) give considerably more detail on many of the topics covered in this chapter.

The Multigrid Tutorial by Briggs, Henson and McCormick (2000) is a useful introduction, while Wesseling (1992) and Hackbusch (1985) analyse multigrid methods in considerable detail.

The conjugate gradient method is discussed in a number of books, such as Golub and Van Loan (1996). More detail is given by van der Vorst (2003) and Elman, Silvester and Wathen (2004). This is an area of vigorous research activity, and these books will help the reader to follow the recent literature.

Exercises

7.1   The 2×2 matrix A is symmetric and positive definite; show that the Jacobi iteration for A converges.

7.2   The matrix A is symmetric and positive definite, and is written in the form $A = L + D + L^T$, where L is strictly lower triangular. Suppose that λ and $\mathbf{z}$ are an eigenvalue and eigenvector of the Gauss–Seidel iteration matrix of A, with $\bar{\mathbf{z}}^T D \mathbf{z} = 1$. Show that

$$\lambda = -\frac{\bar{\mathbf{z}}^T L^T \mathbf{z}}{1 + \bar{\mathbf{z}}^T L \mathbf{z}}.$$

Show also that $2p + 1 > 0$, where p is the real part of $\bar{\mathbf{z}}^T L \mathbf{z}$, and that $|\lambda| < 1$. Deduce that the Gauss–Seidel method converges for any symmetric positive definite matrix.

7.3   The $N \times N$ matrix E has all its elements equal to unity. Show that one of the eigenvalues of E is N, and all the others are zero. Construct a matrix $A = I + kE$, where k is a constant to be determined, such that A is symmetric and positive definite, but in general the Jacobi method diverges. Explain why this result does not contradict Exercise 7.1.

7.4   The system of equations $A\mathbf{u} = \mathbf{b}$ results from use of the standard
         five-point stencil to solve Poisson's equation on the unit square
         with Dirichlet boundary conditions and interval $\Delta x = \Delta y = \frac{1}{3}$.
         The matrix is therefore 4×4. There are 24 such matrices A,
         obtained by numbering the grid points in all possible different
         orders. By considering the symmetries of the problem show
         that there are only three different matrices, each occurring eight
         times. Determine the eigenvalues of the Jacobi iteration matrix,
         and of the three Gauss–Seidel matrices. Verify that for two of
         the matrices the eigenvalues of the Gauss–Seidel matrix are zero
         together with the squares of the eigenvalues of the Jacobi matrix.

7.5   For the solution of $\nabla^2 u + f = 0$ on the unit square we are given
         a Neumann condition on $x = 0$ and Dirichlet conditions on the
         other three sides. The Neumann condition is approximated by
         using an additional line of points corresponding to $x = -\Delta x$
         and using central differences to represent the normal derivative.
         The resulting system of linear equations is then solved by the
         Jacobi iterative method. Show that the errors can be expanded
         in terms of the Fourier modes

$$e_{r,s}^{(n)} = \sum_{k_x, k_y} a^{(n)}(k_x, k_y) \cos k_x x_r \sin k_y y_s$$

         where $k_y = \pi, 2\pi, \ldots, (J-1)\pi$ and $k_x = \frac{1}{2}\pi, \frac{3}{2}\pi, \ldots, (J - \frac{1}{2})\pi$.
         Deduce that the eigenvalues of the Jacobi iteration matrix are

$$\mu_J(k_x, k_y) = \tfrac{1}{2}(\cos k_x \Delta x + \cos k_y \Delta y)$$

         and that for a square mesh with $\Delta x = \Delta y$ the spectral radius
         is approximately $1 - \frac{5}{16}(\pi \Delta x)^2$.

         Extend this analysis to problems where Neumann conditions
         are given on two sides of the square, and then on three sides of
         the square. What happens when Neumann conditions are given
         at all points on the boundary?

7.6   Laplace's equation $u_{xx} + u_{yy} = 0$ is approximated by central
         differences on a square mesh of size $\Delta x = \Delta y = 1/N$ on the
         triangular region defined by $x \geq 0$, $y \geq 0$, $x + y \leq 1$. The
         resulting system of linear equations is solved by the Jacobi iter-
         ative method. Show that the vector with elements

$$w_{r,s} = \sin\left(\frac{pr\pi}{N}\right) \sin\left(\frac{qs\pi}{N}\right) \pm \sin\left(\frac{qr\pi}{N}\right) \sin\left(\frac{ps\pi}{N}\right)$$

is an eigenvector of the iteration matrix. Hence determine the eigenvalues of the iteration matrix, and the asymptotic rate of convergence.

7.7   Determine the von Neumann stability condition for the solution of the equation $u_t = u_{xx} + u_{yy} + f(x, y)$ by the three-term difference scheme

$$\frac{U_{r,s}^{n+1} - U_{r,s}^{n-1}}{2\Delta t} =$$

$$\frac{U_{r,s+1}^n + U_{r,s-1}^n + U_{r+1,s}^n + U_{r-1,s}^n - 2U_{r,s}^{n+1} - 2U_{r,s}^{n-1}}{(\Delta x)^2} + f_{rs}$$

on a uniform square mesh of size $\Delta x = \Delta y$.

Explain how this scheme might be used as an iterative method for the solution of the equations

$$\delta_x^2 U_{r,s} + \delta_y^2 U_{r,s} + (\Delta x)^2 f_{r,s} = 0.$$

Show that taking $\Delta t = \frac{1}{4}(\Delta x)^2$ gives the same rate of convergence as Jacobi's method, but that the choice

$$\Delta t = \frac{1}{4}(\Delta x)^2 / \sin(\pi \Delta x)$$

gives significantly faster convergence.

7.8   Show that the weighted Jacobi method defined by

$$D\mathbf{x}^{(n+1)} = (1 - \omega)D\mathbf{x}^{(n)} + \omega(\mathbf{b} + (L + U)\mathbf{x}^{(n)}).$$

has the iteration matrix

$$G = \left[(1 - \omega)I + \omega D^{-1}(L + U) \right].$$

Show that for the problem of Section 6.1 the eigenvalues of this matrix are

$$\mu_{rs} = 1 - \omega \left(\sin^2 \frac{r\pi}{2J} + \sin^2 \frac{s\pi}{2J} \right),$$

for $r, s = 1, 2, \ldots, J - 1$. Defining the smooth eigenvectors to be those for which $2r \leq J$ and $2s \leq J$, verify that the eigenvalues corresponding to the nonsmooth eigenvectors lie between $1 - \frac{1}{2}\omega$ and $1 - 2\omega$. Deduce that the best performance as a smoother is obtained by the choice $\omega = 2/5$, and that with this choice every nonsmooth component of the error is reduced by a factor at least $3/5$.

7.9   Repeat the previous example for the three dimensional problem of the solution of Poisson's equation in the unit cube, with
Dirichlet boundary conditions. Find the optimum choice of ω,
and show that weighted Jacobi iteration now reduces each nonsmooth component of the error by a factor of at least $5/7$.

7.10   In the conjugate gradient iteration as defined by (7.61) with
$M = I$, the unit matrix, show by induction that, for each posi-
tive integer n, the vectors $\mathbf{x}^{(n)}$, $\mathbf{p}^{(n-1)}$ and $\mathbf{r}^{(n-1)}$ all lie in the
Krylov space K^n, which is the space spanned by the vectors
$\mathbf{b}, A\mathbf{b}, \ldots, A^{n-1}\mathbf{b}$.

Now suppose that the $N \times N$ matrix A has only k distinct
eigenvalues, where $k \leq N$. Show that the dimension of the space
K^n does not exceed k, and deduce that the iteration terminates,
with the exact solution of $A\mathbf{x} = \mathbf{b}$, in at most $k + 1$ iterations.

7.11   The conjugate gradient algorithm described in (7.61) begins
with the first approximation $\mathbf{x}^{(0)} = 0$. Suppose that an approximation $\mathbf{x}^*$ to the solution is available: suggest how the method
might be modified to make use of this information.

References

[Compiled from the footnotes and bibliographic notes in each chapter.]

Ames, W. F. (1965). *Nonlinear Partial Differential Equations in Engineering,* Vol. I. New York, Academic Press.

— (1972). *Nonlinear Partial Differential Equations in Engineering,* Vol. II. New York, Academic Press.

— (1992). *Numerical Methods for Partial Differential Equations,* 3rd edn. Boston, Academic Press.

Brandt, A. (1977). Multilevel adaptive solutions to boundary-value problems. *Math. Comp.* **31**, 333–90.

Brenner, S. and Scott, L. R. (2002). *The Mathematical Theory of Finite Element Methods,* Second Edition. New York, Springer.

Bridges, T. J. (1997). Multi-symplectic structure and wave propagation. *Math. Proc. Camb. Philos. Soc.* **121**, 147–90.

Briggs, W. L., Henson, V. E. and McCormick, S. F. (2000). *A Multigrid Tutorial,* Second edition. SIAM.

Buchanan, M. L. (1963a). A necessary and sufficient condition for stability of difference schemes for initial value problems. *J. Soc. Indust. Appl. Math.* **11**, 919–35.

— (1963b). A necessary and sufficient condition for stability of difference schemes for second order initial value problems. *J. Soc. Indust. Appl. Math.* **11**, 474–501.

Carrier, G. F. and Pearson, C. E. (1976). *Partial Differential Equations.* New York, Academic Press.

Ciarlet, P. G. (1978). *The Finite Element Method for Elliptic Problems.* North Holland, Amsterdam.

Cockburn, B. (2003). Continuous dependence and error estimation for viscosity methods. *Acta Numerica* **12**, 127–80.

Colella, P. and Woodward, P. R. (1984). The piecewise parabolic method (ppm) for gas-dynamical simulations. *J. of Comput. Phys.* **54**, 174–201.

Collatz, L. O. (1966). *The Numerical Treatment of Differential Equations.* Berlin, Springer.

Courant, R. and Friedrichs, K. O. (1948). *Supersonic flow and shock waves.* New York, Wiley-Interscience.

Courant, R., Friedrichs, K. O. and Lewy, H. (1928). Über die partiellen Differenzengleichungen der mathematischen Physik. *Math. Ann.* **100**, 32–74.

Courant, R. and Hilbert, D. (1962). *Methods of Mathematical Physics, Vol 2: Partial Differential Equations.* New York, Wiley-Interscience.

Crank, J. (1975). *The Mathematics of Diffusion,* 2nd edn. Oxford, Clarendon Press.

Crank, J. and Nicolson, P. (1947). A practical method for numerical evaluation of solutions of partial differential equations of the heat-conduction type. *Proc. Camb. Philos. Soc.* **43**, 50–67.

Douglas, J. Jr and Rachford, H. H. Jr (1956). On the numerical solution of the heat conduction problems in two and three variables. *Trans. Amer. Math. Soc.* **82**, 421–39.

Duff, I. S., Erisman, A. M. and Reid, J. K. (1986). *Direct Methods for Sparse Matrices.* Oxford, Clarendon Press.

D'yakonov, E. G. (1964). Difference schemes of second order accuracy with a splitting operator for parabolic equations without mixed partial derivatives. *Zh. Vychisl. Mat. i Mat. Fiz., Moscow* **4**, 935–41.

Elman, H. C., Silvester, D. J. and Wathen, A. J. (2004). *Finite Elements and Fast Iterative Solvers.* Oxford, Oxford University Press.

Engquist, B. and Osher, O. (1981). One-sided difference approximations for nonlinear conservation laws. *Math. Comp.* **36**, 321–52.

Evans, L. C. (1998). *Partial Differential Equations,* Graduate Studies in Mathematics. Providence, Rhode Island, American Mathematical Society.

Godunov, S. K. (1959). A finite difference method for the numerical computation of discontinuous solutions of the equations of fluid dynamics. *Mat. Sb.* **47**, 271–306.

Godunov, S. K. and Ryabenkii, V. S. (1964). *The Theory of Difference Schemes - An Introduction.* North Holland, Amsterdam.

Golub, G. H. and Van Loan, C. F. (1996). *Matrix Computations,* 3rd edn. Baltimore, Johns Hopkins University Press.

Gottlieb, S., Shu, C. W. and Tadmor, E. (2001). Strong stability-preserving high-order time discretisation methods. *SIAM Rev.* **43**, 89–112.

Gustaffson, B., Kreiss, H. O. and Sundström, A. (1972). Stability theory of difference approximations for mixed initial boundary value problems ii. *Math. Comp.* **24**(119), 649–86.

Hackbusch, W. (1985). *Multigrid methods and applications.* Berlin, Springer-Verlag.

(1994). *Iterative solution of large sparse linear systems of equations.* Berlin, Springer-Verlag.

Hairer, E., Lubich, C. and Wanner, G. (2002). *Geometric Numerical Integration.* Berlin, Springer-Verlag.

Harten, A. (1984). On a class of high resolution total-variation-stable finite-difference schemes. *SIAM J. Numer. Anal.* **21**, 1–23.

Hestenes, M. R. and Stiefel, E. (1952). Methods of conjugate gradients for solving linear systems. *J. Res. Nat. Bur. Stand.* **49**, 409–36.

John, F. (1952). On the integration of parabolic equations by difference methods. *Comm. Pure Appl. Math.* **5**, 155.

Keller, H. B. (1971). A new finite difference scheme for parabolic problems. In B. Hubbard (ed.), *Numerical Solution of Partial Differential Equations II, SYNSPADE 1970,* Academic Press, pp. 327–50.

Kraaijevanger, J. F. B. M. (1992). Maximum norm contractivity of discretization schemes for the heat equation. *Appl. Numer. Math.* **99**, 475–92.

Kreiss, H. O. (1962). Über die Stabilitätsdefinition für Differenzengleichungen die partielle Differentialgleichungen approximieren. *Nordisk Tidskr. Informations-Behandling* **2**, 153–81.

(1964). On difference approximations of the dissipative type for hyperbolic differential equations. *Comm. Pure Appl. Math.* **17**, 335–53.

Kreiss, H. O. and Lorenz, J. (1989). *Initial-Boundary Value Problems and the Navier–Stokes Equations.* Academic Press, San Diego.

Lax, P. D. and Richtmyer, R. D. (1956). Survey of the stability of linear finite difference equations. *Comm. Pure Appl. Math.* **9**, 267–93.

Lax, P. D. and Wendroff, B. (1960). Systems of conservation laws. *Comm. Pure and Appl. Math.* **13**, 217–37.

Leimkuhler, B. and Reich, S. (2004). *Simulating Hamiltonian Dynamics.* Cambridge, Cambridge University Press.

LeVeque, R. J. (1992). *Numerical Methods for Conservation Laws,* Lectures in Mathematics ETH Zurich, 2nd edn. Basel, Birkhauser Verlag.

(2002). Finite Volume Methods for Hyperbolic Problems. Cambridge, Cambridge University Press.

LeVeque, R. J. and Trefethen, L. N. (1988). Fourier analysis of the SOR iteration. *IMA J. Numer. Anal.* **8(3)**, 273–9.

Levy, D. and Tadmor, E. (1998). From semi-discrete to fully discrete: stability of Runge–Kutta schemes by the energy method. *SIAM Rev.* **40**, 40–73.

Lighthill, M.J. (1978). *Waves in Fluids.* Cambridge, Cambridge University Press.

López-Marcos, J. C. and Sanz-Serna, J. M. (1988). Stability and convergence in numerical analysis iii: linear investigation of nonlinear stability. *IMA J. Numer. Anal.* **8(1)**, 71–84.

Matsuno, T. (1966). False reflection of waves at the boundary due to the use of finite differences. *J. Meteorol. Soc. Japan* **44(2)**, 145–57.

Miller, J. (1971). On the location of zeros of certain classes of polynomials with applications to numerical analysis. *J. Inst. Math. Appl.* **8**, 397–406.

Mitchell, A. R. and Griffiths, D. F. (1980). *The Finite Difference Method in Partial Differential Equations.* New York, Wiley-Interscience.

Morton, K. W. (1996). *Numerical Solution of Convection–Diffusion Problems.* London, Chapman & Hall.

Peaceman, D. W. and Rachford, H. H. Jr (1955). The numerical solution of parabolic and elliptic differential equations. *J. Soc. Indust. Appl. Math.* **3**, 28–41.

Preissmann, A. (1961). *Propagation des intumescences dans les canaux et rivières,* paper presented at the First Congress of the French Association for Computation, Grenoble, France.

Protter, M. H. and Weinberger, H. F. (1967). *Maximum Principles in Differential Equations.* Englewood Cliffs, NJ, Prentice-Hall.

Reich, S. (1999). Backward error analysis for numerical integration. *SIAM J. Numer. Anal.* **36**, 1549–1570.

Richtmyer, R. D. and Morton, K. W. (1967). *Difference Methods for Initial Value Problems,* 2nd edn. New York, Wiley-Interscience. Reprinted (1994), New York, Kreiger.

Roe, P. L. (1981). Approximate Riemann solvers, parameter vectors, and difference schemes. *J. of Comput. Phys.* **43**, 357–72.

Roos, H. G., Stynes, M. and Tobiska, L. (1996). *Numerical Methods for Singularly Perturbed Differential Equations.* Berlin, Springer.

Saad, Y. and Schultz, M. H. (1986). A generalised minimal residual algorithm for solving nonsymmetric linear systems. *J. Sci. Stat. Comput.* **7**, 856–69.

Shokin, Y. I. (1983). *The Method of Differential Approximation.* New York–Berlin, Springer.

Smoller, J. (1983). *Shock Waves and Reaction-Diffusion Equations.* New York, Springer-Verlag.

Strang, G. (1964). Wiener–Hopf difference equations. *J. Math. Mech.* **13(1)**, 85–96.

Strang, G. and Fix, G. (1973). *An Analysis of the Finite Element Method.* New York, Prentice-Hall.

Süli, E. and Mayers, D. F. (2003). *An Introduction to Numerical Analysis.* Cambridge, Cambridge University Press.

Tadmor, E. (2003). Entropy stability theory for difference approximations of nonlinear conservation laws and related time-dependent problems. *Acta Numerica* **12**, 451–512.

Thomée, V. (1962). A stable difference scheme for the mixed boundary value problem for a hyperbolic first order system in two dimensions. *J. Soc. Indust. Appl. Math.* **10**, 229–45.

Trefethen, L. N. (1982). Group velocity in finite difference schemes. *SIAM Rev.* **24**, 113–36.

van der Vorst, H. A. (2003). *Iterative Krylov Methods for Large Linear Systems.* Cambridge, Cambridge University Press.

van Leer, B. (1974). Towards the ultimate conservative difference scheme. II: monotonicity and conservation combined in a second order scheme. *J. of Comput. Phys.* **14**, 361–70.

(1979). Towards the ultimate conservative difference scheme V. a second order sequel to Godunov's method. *J. Comput. Phys.* **32**, 101–36.

Varga, R. S. (1974). *Matrix Iterative Analysis.* Englewood Cliffs, NJ, Prentice-Hall.

Warming, R. F. and Hyett, B. J. (1974). The modified equation approach to the stability and accuracy of finite difference methods. *J. of Comput. Phys.* **14**, 159–79.

Wesseling, P. (1992). *An Introduction to Multigrid Methods.* Chichester, John Wiley.

Whitham, G. B. (1974). *Linear and Nonlinear Waves.* New York, Wiley-Interscience.

Yanenko, N. N. (1971). *The Method of Fractional Steps*. Berlin, Springer-Verlag.

Yanenko, N. N. and Shokin, Y. I. (1969). First differential approximation method and approximate viscosity of difference schemes. *Phys. of Fluids* **12**, Suppl. II, 28–33.

Young, D. M. (1971). *Iterative Solutions of Large Linear Systems*. New York, Academic Press.

Zhao, P. F. and Quin, M. Z. (2000). Multisymplectic geometry and multisymplectic Preissmann scheme for the kdv equation. *J. Phys. A* **33**, 3613–26.

Index

Printed in the United States
By Bookmasters